CHASING LAVA

A Geologist's Adventures at the Hawaiian Volcano Observatory

Wendell A. Duffield

Published in cooperation with the U.S. Geological Survey's
Volcano Hazards Program

Mountain Press Publishing Company
Missoula, Montana
2003

Studies carried out under the U.S. Geological Survey's Volcano Hazards Program advance scientific understanding of volcanic processes and help reduce harmful eruption impacts on people and property.

Front cover: Two-thousand-foot-tall eruption at Mauna Ulu —*Wendell Duffield photo, U.S. Geological Survey*

Back cover: A lava fountain from Mauna Ulu —*Wendell Duffield photo, U.S. Geological Survey*

"A Note About the Hawaiian Language" is reprinted, with permission from the publisher, from *Roadside Geology of Hawai'i,* by Richard W. Hazlett and Donald W. Hyndman, Mountain Press Publishing Company, 1996.

All illustrations by Sara Boore and Susan Mayfield unless otherwise credited.

Library of Congress Cataloging-in-Publication Data

Duffield, Wendell A.
Chasing lava : a geologist's adventures at the Hawaiian Volcano Observatory / Wendell A. Duffield.
p. cm.
Includes bibliographical references and index.
ISBN 0-87842-462-8 (pbk. : alk. paper)
1. Kilauea Volcano (Hawaii) 2. Volcanism—Hawaii—Kilauea Volcano.
I. Title.
QE523.K5 D84 2003
551.21'09969'1—dc21

2002011273

Printed in Hong Kong by Mantec Production Company

Mountain Press Publishing Company
P.O. Box 2399 • Missoula, Montana 59806
406-728-1900

To all the HVO staff, present and past, recruited from the local Hawaiian work force. These loyal and devoted employees provide continuity necessary to sustain a high level of productivity at the observatory. I especially recognize Reggie Okamura, with whom I worked closely during my three years at HVO but who died just a little too soon to read the stories I've written about our joint adventures.

I also dedicate the book to Richard V. Fisher, who was an extraordinary volcanologist, professor, and friend. He was a role model for many earth scientists and an inspiration to those who dare to try to explain their science to the general public.

CONTENTS

ACKNOWLEDGMENTS

This book experienced an elephantine-scale gestation period. I frequently thought about writing it, from the moment I left the Hawaiian Volcano Observatory in 1972, but other demands on my time precluded any early action. My retirement in 1997, following a long career with the U.S. Geological Survey, finally presented the opportunity to write.

The roughly twenty-five-year lag between events and their telling carries both disadvantages and advantages. On the down side, the memory of past events tends to lose accuracy with time. Still, I have written my stories as truthfully as my memory and notes allow. I have also enlisted the memories of several friends who were at HVO with me. The passage of time has provided me with the benefit of a more mature and engaging perspective than I had as a sometimes hotheaded volcanologist in his early thirties.

Win Banko, Dallas Jackson, Jeffrey Judd, Arnold Okamura, Don Peterson, and Don Swanson have served as critics and memory checkers, especially for the parts of the book that include them as characters. The two Dons also critically read the entire manuscript. George Ulrich, a man who has felt the heat, provided additional insights for the chapter about human legs immersed in active lava flows. I greatly appreciate the help of all these friends, and I accept the blame for any mistakes that may still exist in the book. Bob Tilling, a highly valued mentor, sometimes supervisor, and long-time professional colleague, provided many helpful comments on early versions of the text and was the fundamental force behind the U.S. Geological Survey's efforts to successfully forge a publication alliance with Mountain Press. Within Mountain Press, my editor, Kathleen Ort, provided encouragement, direction, and the overall push to get the book into print. Sara Boore and Susan Mayfield of the U.S. Geological Survey developed the book's design and final versions of diagrams and sketches.

Finally, I praise the tolerance and patience of my wife, Anne. She has permitted me to intrude on her privacy by making her a character in the book, and she endured (or enjoyed?) many hours and days of my absence from home as I sat composing at the office keyboard.

A NOTE ABOUT HAWAIIAN LANGUAGE

The following guidelines will assist readers who are unfamiliar with how to pronounce Hawaiian words. Clusters of vowel combinations and a small number of consonants characterize the Hawaiian language. Two diacritical marks used in Hawaiian affect pronunciation. They also affect a word's meaning—for example, lanai (stiff-backed chair) and lānai (porch or verandah), and ono (fish) and ʻono (delicious).

Hawaiian has eight consonants (h, k, l, m, n, p, w, ʻ). The first seven are pronounced roughly the same as in English. The sound represented by an okina, a single opening quotation mark (ʻ), is a glottal stop, pronounced like the sound in "oh-oh."

When you know how to pronounce Hawaiian vowels, pronouncing Hawaiian words is not difficult. Two vowels, *e* and *i*, are pronounced differently than they usually are in English: *e* sounds close to a long *a*, and *i* sounds like a long *e*. Vowels with a *kahako* or macron—a bar—over them are pronounced with greater than normal stress and duration. Pronounce Hawaiian vowels as follows:

a as in l*a*va	m*a*h*a*lo
e as in b*e*t	h*e*l*e* (go)
ē as in h*a*y	n*ē*n*ē* (Hawaiian goose)
i as in f*ee*t	Waik*ī*k*i*
o as in h*o*le	al*o*ha
u as in b*oo*t	h*u*la

Diphthongs produce various vowel sounds: *au* as in Maui, *oi* as in *poi*. A long *i* sound comes from the diphthong *ai*, as in *makai* (toward the sea).

Determining where to put the stress in a Hawaiian word can be baffling for someone not used to hearing Hawaiian words pronounced. For shorter words, the next-to-last syllable is often stressed (wa*hi*ne, a*lo*ha, Oʻ*a*hu). Also, a syllable with a long vowel or a diphthong—*lā* (sun), *kai* (sea)—is often stressed.

1

A FARM BOY'S IMAGE OF HAWAI'I

Hawaiʻi. For many people from the Upper Midwest, where my wife Anne and I grew up, this word alone conjures up thoughts of an idyllic tropical paradise. The mind's eye sees vast expanses of sandy beach washed by crystal-clear water—water warm enough to enter easily without the fear of a July "Minnesota-lake chill," yet not so warm to induce a saunalike lethargy. Wintry visions of heavy coats and wool-lined boots fade to images of bare-the-body weather year-round and the occasional rain that freshens the landscape and the human spirit alike. Memories of the heavy scent and monotonous taste of meat, potatoes, and Upper Midwest delicacies known as hot dish and casserole give way to the delicate textures and flavors of abundant fresh seafood and a seemingly unending in-season crop of sweet fruit. Visions of orderly rows of corn and soybean plants marching across a flat landscape in cardinal compass directions blur to chaotic mountains adorned with colorful profusions of exotic plants that Midwesterners see only in TV documentaries. The Hawaiian dream also generally includes the idea of partying late at night and sleeping in equally late each morning, without the fear of a wakeup call from the boss or surges of Protestant-work-ethic guilt. After all, to most mainlanders, Hawai'i means vacation.

Anne and I were lucky enough to enjoy life in Hawai'i for three years, from 1969 to 1972. The occasion was not an extra-long vacation, although many a friend and professional colleague have suggested otherwise. Appearances and popular conceptions notwithstanding, I was temporarily transferred to Hawai'i for work, as a regular part of my job. The U.S. Geological Survey sent me, a fledgling geologist just recently graduated from Stanford University, to help staff the Hawaiian Volcano Observatory, a global outpost for earth scientists since 1912. I was sent there to help learn about the life of a typical Hawaiian volcano, if indeed there is such a creature, not to learn about and enjoy the good human life in Hawai'i. By simply being there, though, I could not avoid experiencing the pleasures of the tropics—and experience them we did. But I spent most of my time charting and trying to interpret the vagaries of the behavior of Kīlauea Volcano.

Some describe Kīlauea as a "drive-in" volcano; others call it the world's "most active" volcano. The former label is accurate and has been ever since the introduction of motor vehicles to Hawai'i in the early part of the twentieth century. Even before then, Kīlauea was very accessible by foot and by horse because this volcano was, and still is, central to Hawaiian culture as well as a popular tourist destination. The "most-active" label is arguable but probably is important only for local bragging rights anyway. The fundamentally important fact is that Kīlauea erupts frequently relative to a typical human life span. The volcano's eruptive liveliness and easy accessibility make it an excellent place for geologists to study the why, when, where, and how of volcanism.

Now, about ninety years after the Hawaiian Volcano Observatory opened its doors, geologists have learned to understand many aspects of Kīlauea's behavior. However, the volcano has yielded this knowledge unevenly, perhaps grudgingly; if Kīlauea holds one universal truth for the probing scientist, it is to maintain a healthy level of humility. More than once, Kīlauea has leaked important new bits of information just when "modern" scientists thought they had finally pieced together the volcano's complete story. As Primo Levi wrote, "There is trouble in store for anyone who surrenders to the temptation of mistaking an elegant hypothesis for certainty."

The life spans of many volcanoes greatly exceed that of any human. And each volcano is unique in some way. It takes the combined efforts of many generations of geologists to sample even a moderate fraction of any volcano's cradle-to-grave story. The task for volcanologists may seem impossible. Impossible or not, my three-year stint at the Hawaiian Volcano Observatory was one of the most fascinating phases of my life, both professionally and personally. The following collection of essays recounts highlights from those years. Some of the stories are only indirectly related to Kīlauea, the active volcano. However, all are about Kīlauea, the place, which would be fascinating even without the occasional shaking and breaking of ground and belching of molten rock. My hope is that you can take away a better understanding of Hawaiian volcanoes as well as enjoyment from these tales about two Lake Wobegon kids temporarily transplanted to a tropical paradise.

An active pāhoehoe lava flow envelops a small (6 feet tall) ʻōhiʻa tree on Kīlauea.
—U.S. Geological Survey photo

2 A VOLCANO OBSERVATORY IS BORN

Emboldened with the wisdom that accompanies hindsight, I can confidently tell you that much of my adult life, both personal and professional, was predestined by events that occurred long before I was born. At the time my die was cast, my parents were only children themselves, living hundreds of miles apart in Middle America. My destiny jelled in January 1912 on an island near the middle of the Pacific Ocean. The event was the formal establishment of a volcano observatory at the summit of Kīlauea Volcano on the island of Hawai'i. And the orchestrator was Thomas Augustus Jaggar, an earth scientist from the Massachusetts Institute of Technology.

Though not the first of its kind, the Hawaiian Volcano Observatory, or HVO, became one of three widely recognized volcano observatories shortly after it came into being. The Japanese had established a similar operation at Asama Volcano the year before, and concerned Italians had been closely watching the behavior of Vesuvius for sixty-five years, an effort that the A.D. 79 residents of nearby Pompeii and Herculaneum might have wished for. Still, many considered HVO one of the world's premier volcano observatories back in 1912, and it is arguably the most prestigious today.

The birth of the Hawaiian Volcano Observatory marked the beginning of scientists' long-term observations of active Hawaiian volcanoes in an attempt to learn how they behave and to apply this knowledge to defining the range of and limits to "safe" interactions between people and active volcanoes. It was a very ambitious undertaking by an extremely talented and energetic man. Jaggar could not have guessed how much his actions would influence the life of this as-yet-unborn farm boy.

Certainly, many volcanoes were "observed" closely elsewhere around the world centuries, if not millennia, ago; people who live on or very near an active volcano are naturally concerned about the behavior of their powerful and potentially lethal neighbor. Moreover, volcano observatories exist today in such diverse places as France (Réunion and Caribbean Islands), Indonesia, Italy, Japan, New Zealand, Philippines, Russia, and the mainland United States. Nonetheless, HVO's written and oral history since 1912 comprises an enviable library of accomplishments that began with the birth of this volcano observatory under the guidance, wisdom, and creativity of Jaggar.

Industrialization was well under way in many parts of the world during the nineteenth and early twentieth centuries, and with concomitant growth of human population, the frequency of unfortunate and sometimes fatal interactions between people and volcanoes similarly grew. In 1902, the eruption of Mt. Pelée on the island of Martinique in the Caribbean Sea sent a searing-hot hurricane of volcanic ash, gas, and lava howling down the mountain's flank, destroying the nearby town of St. Pierre and simultaneously killing nearly all the town's 28,000 inhabitants. The eruption's toll in human lives profoundly disturbed Jaggar and many of his contemporary earth scientists. History records several other volcanic eruptions that wrought mass human deaths, and Jaggar realized the future would likely bring more such disasters as people increasingly chose to live on or near active volcanoes.

Rather than simply watch with academic interest as the situation became worse, Jaggar and his like-minded contemporaries chose to act. The eruption of Mt. Pelée and its aftermath catalyzed Jaggar's belief that

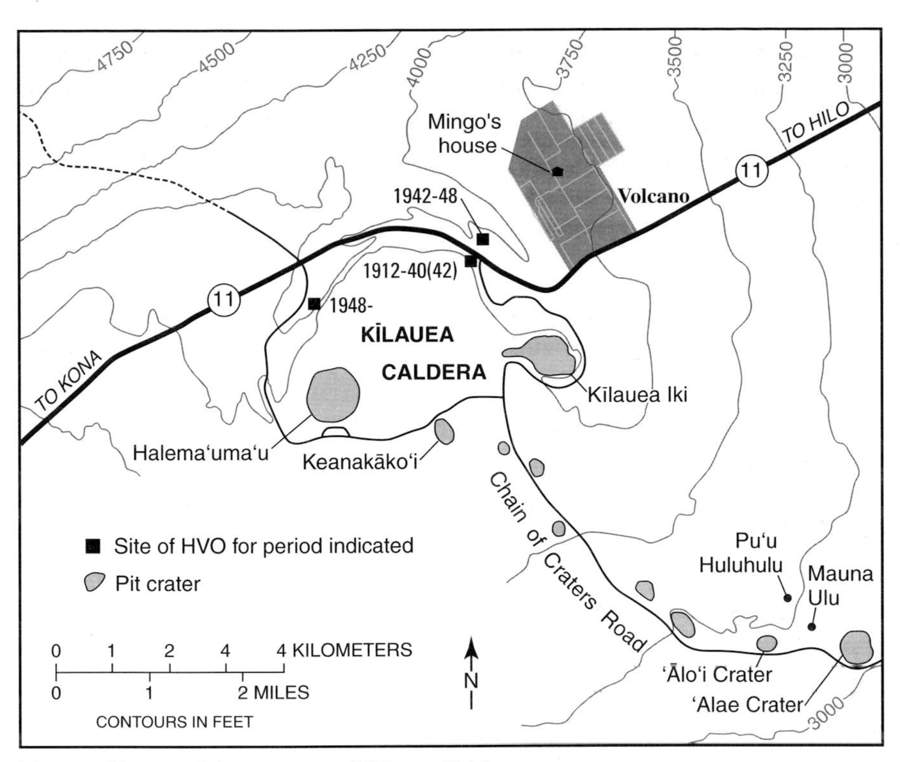

Topographic map of the area around Kīlauea Caldera

systematic and long-term studies of volcanoes were sorely needed. As Jaggar saw it, the key to minimizing fatal interactions between people and volcanoes was to understand volcanic behavior and to apply this knowledge to the planning of future human activities near active volcanoes. Within a decade, he established his fledgling volcano observatory and research center, HVO, on Kīlauea Volcano in the middle of the Pacific Ocean, halfway around the world from Mount Pelée. The observatory's motto reflected Jaggar's primary research goal: *Ne plus haustae aut obrutae urbes* ("No more burned or buried cities"). As Jaggar wrote, "The attitude of geologists has never accorded with mine. They are trying to explain ancient structures. I am trying to explain future events."

Why establish one of the world's first serious volcano observatories on Hawai'i rather than Martinique? The answer is twofold. First, volcanoes like Mount Pelée are built from a type of lava that is so thick and pasty that the volcanoes' flanks are too steep and rugged for easy human access. This has the practical consequences of making field studies difficult, if not impossible. By contrast, the volcanoes of Hawai'i are built from a thin, runny type of lava that produces a mountain with gentle, readily accessible slopes. Thus, easy-to-construct roadways crisscross Kīlauea Volcano, and people can reach the hinterlands of the volcano on foot or pack animal.

More important, though, Hawaiian volcanoes are of a type that erupt many times during a typical human lifetime, whereas Mount Pelée and its Caribbean volcano neighbors typically erupt only once every century or so. Imagine the folly of building an observatory and then having to wait through many human generations for the first eruption that scientists could conveniently study from that location.

HVO follows a policy of rotational research staffing, meaning that most scientists serve an average of only four years there. (Such rotation exposes a practical maximum number of scientists to training, discovery, and intellectual growth while the observatory simultaneously benefits from a variety of skills and personalities.) Yet during HVO's ninety or so years of existence, very few scientists have spent their entire observatory tour without experiencing at least one eruption. Jaggar recognized that researchers could quickly amass a rich and robust body of data about eruptive behavior at Kīlauea, thus affording the staff an immediate opportunity to analyze and interpret the information and recommend ways to mitigate volcanic hazards.

As Jaggar's awareness of and concern for volcanic effects on people developed, a parallel awakening was under way in Hawai'i. All Hawaiian Islands are of volcanic origin, and as residents are often rudely reminded, the Big Island of Hawai'i is still growing with each eruption from Hualālai, Mauna Loa, and Kīlauea Volcanoes. Humans, beginning with the arrival of the first Christian missionaries in the 1820s, have systematically chronicled the history of eruptions on the Big Island. A combination of provincial pride and awestruck respect for these molten-rock-belching giants grew within the local population. Famous scientists and journalists visited the island to gather information that became articles in learned scientific journals and popular publications read around the world. Thus, even before Jaggar built his volcano observatory, a group of actively interested Hawaiian citizens stood ready to help launch a new volcano research center. But the challenges were many.

To establish a volcano observatory requires both people and money. Fortunately, Jaggar came with enough of both to get under way. His operation initially received support from a research endowment, the Whitney

Fund, administered through his home academic institution, the Massachusetts Institute of Technology. The Hawaiian Volcano Research Association, a group of Hawaiian businessmen led by a volcanophile named Lorrin Thurston, supplemented this fund. The combined resources were sufficient to pay for construction of a modest observatory building, called

The first buildings of the Hawaiian Volcano Observatory, viewed to the north-northwest in this photograph from 1922. Ground level served as Jaggar's office and laboratory. The cellar housed a seismometer. An adjacent machine shop is to the right. Two water tanks next to vehicle store rain runoff from the roof. Volcano House Hotel and a guest cottage are in the right background. Kīlauea Caldera is just out of view to the left. —Courtesy of Bishop Museum

the Whitney Laboratory of Seismology, along the eastern rim of Kīlauea Caldera. Jaggar's office was at ground level. At this initial observatory, the cellar, called the Whitney Vault, housed a seismometer to record earthquakes.

The two World Wars disrupted the continuity in staffing, and vicissitudes in funding passed sponsorship of HVO from organization to organization during its first three-and-one-half decades. From 1912 to 1919,

Jaggar at work in his HVO office —Courtesy of Bishop Museum

support came from the combined resources of the original sponsors, the Whitney Fund and the Hawaiian Volcano Research Association. However, eventual loss of support from the Whitney Fund led the United States Weather Bureau to run the observatory from 1919 to 1924. Next, the U.S. Geological Survey managed and supported HVO until 1935, when the agency passed the sponsorship baton to the National Park Service.

Park Service sponsorship seemed logical—after all, Kīlauea Volcano lies within Hawai'i Volcanoes National Park, which was established in 1916 as Hawai'i National Park. However, support to operate the volcano research center, or any aspect of the new park, often fell victim to an early-twentieth-century political attitude, which a congressman summed up by saying, "It should not cost anything to run a volcano." But, as Jaggar pointed out, "Hawai'i National Park differs from Sequoia in that its big trees are scientific events in time, and it was established by Congress for scientific observatory values, and scientific assistance to endangered populations."

Politics aside, the mandate and staffing of the U.S. Geological Survey were a better match for the observatory's research and the application of the resulting knowledge. HVO was passed back to the U.S. Geological Survey in 1948, where it has remained since. That same year, HVO's daily operations were relocated to a building at Uwēkahuna Bluff, directly

Jaggar in the seismometer cellar—Courtesy of Bishop Museum

across the caldera from the original Whitney Laboratory. Jaggar's ground-level office in the Whitney building was razed to make room for a luxurious lodge, the renovated and enlarged Volcano House Hotel, which is the principal overnight tourist destination for park visitors. The Whitney building's basement, where the seismometer resided, became an empty and incidental room attached to the cellar of the hotel. A grass-covered, 6-foot-tall dome on the caldera-side lawn of the hotel marks the site of the former Whitney Lab. I cannot stroll past that spot without thinking of Jaggar and his fledgling observatory.

Through nearly all the organizational disruptions, Jaggar led HVO and established a scientific framework for a host of studies that were to follow his eventual retirement in 1940. Developing an understanding of Kīlauea Volcano meant charting local earthquakes, measuring and mapping obvious bulges and sinks in and around the volcano, and learning about the properties of magma. With these objectives in mind, Jaggar set to work observing and measuring a lava lake at Halemaʻumaʻu Crater, conveniently located within Kīlauea Caldera, just minutes from his office.

As one of his earliest projects, Jaggar measured the temperature of the molten lava that filled the lake. This task may seem trivial, but Jaggar

Jaggar (holding the steel pole) *and helpers prepare to measure the temperature of the lava in Halemaʻumaʻu by submersing various mixtures of salt and clay, in the cylinder attached to the steel pole, into the lava lake. Lorrin Thurston is immediately behind Jaggar.*
—Courtesy of Bishop Museum

quickly confirmed the difficulties of working in a medium that is nearly the temperature of molten steel. Unlike pig iron in a mold, this hot lava did not necessarily sit passively in its Halemaʻumaʻu home. The lava lake bubbled, splashed, and even overflowed the rim of its crater, keeping Jaggar and other researchers wary and ready to run to higher ground. And Jaggar's initial attempts to lower a thermometer into the lava failed for mechanical reasons: lack of an instrument that could consistently penetrate the semirigid crust that veneers all but the most active of lava lakes. Jaggar tried forcibly inserting a steel pipe that contained several different mixtures of salt and clay, each of a known and different melting temperature, a primitive sort of thermometer. But this technique produced temperature readings that were clearly too low, likely because Jaggar and his helpers couldn't keep the pipe in the magma long enough for it to reach the temperature of the lava.

Having inserted various sorts of probes into molten lava myself, I can readily appreciate the difficulties and frustrations Jaggar and his crew experienced. Even under the cover of protective clothing, heat radiating from the lava quickly drives one into retreat for cooler air. Nonetheless, Jaggar's failure to gather accurate lava temperatures succeeded in defining some physical and logistical obstacles to inserting instruments into that hot and chemically corrosive substance.

Jaggar's studies of volcano-related earthquakes were more fruitful. Through trial and error, he learned to modify a conventional seismometer to respond to the frequent ground shaking that originates directly underfoot at Kīlauea, in the magma-saturated bowels of the volcano. Before long, he began to recognize patterns and relationships between the timing of earthquakes and volcanic eruptions. It seemed that magma announced its subterranean movements by shaking and breaking adjacent rocks, presumably creating conduits through which the magma could flow and erupt.

While modifying his seismometer to clearly record magma-generated earthquakes, Jaggar discovered that the pendulum, a fundamental component of the instrument, was also recording changes in the slope, or tilt, of the ground. To function properly, the base of his seismometer had to be horizontal, with the tug of gravity keeping its pendulum vertical. Yet, repeatedly, over days or even just hours, the pendulum of the seismometer in the Whitney Vault went out of plumb relative to the instrument's base. The ground under the vault apparently was being tilted.

The need to frequently readjust the seismometer was at once an aggravation and an independent source of information about the volcano. The volcano's surface tilted in response to the subterranean movement of magma. Magma was pretty clearly pushing up into the volcano from below, bulging the earth's surface somewhere west of, yet near, the seismometer in the Whitney Vault. Jaggar understood that he could precisely locate the center of this bulge if it were surrounded by tiltmeters, but such an ambitious network of instruments would have to come later, with more funding and a larger staff. Meanwhile, repeated surveys to determine the elevations of reference markers around the volcano's summit helped show just where the bulging focused.

Measuring the elevation of the ground relative to sea level is a standard part of many land surveys and a requirement for constructing a contour map. Such a map was a very early product of Jaggar's research program. When they repeated land-elevation surveys and compared the measurements to those on the original map, Jaggar and his surveyors discovered that the summit area of Kīlauea moves, sometimes up and sometimes down. Not surprisingly, bulging was precursory to eruption, as new magma moved from great depth up into the volcano itself. Subsidence came with or even a bit after eruption, as the magma-engorged volcano bled off some of its hot bloat. Even though the researchers might have expected such ups and downs, the newly acquired surveying data proved the existence, size, and location of the bulges for the first time. This was pioneering stuff, truly pushing back the frontiers of science. One more set of surveying data added power to this push.

While leveling determines the elevation, or vertical position, of the ground, a surveying technique called triangulation defines the horizontal position. As the name implies, reference marks on the ground serve as the corners of a network of triangles. The surveyors define the positions of these marks by precisely measuring the internal angles of each triangle. Repeated triangulation indicated that the internal angles were changing—the ground at Kīlauea was moving laterally by varying amounts, as well as vertically. Substantial groundwork was now in place for a grand synthesis of how the volcano responds to the movements of magma into and through the mountain.

From the measurements of ground deformation and the record of volcano-generated earthquakes, Jaggar "discovered" that the summit of Kīlauea swells and shakes as magma rises into the mountain from its mantle source and then subsides with continued earthquakes as it loses magma through eruption—inflating and deflating much like a balloon. Jaggar's balloon analogy persists in only slightly modified form today. The many limitations of instrumentation continually plagued Jaggar's research program and nagged at the certainty of his model of volcano behavior. But beyond Jaggar's tenure at HVO, large networks of increasingly reliable and sensitive instruments substantiated his initial interpretation.

Jaggar accomplished a variety of other firsts during his nearly three decades at HVO. During times of relative quiet at Kīlauea, he turned his attention to the much larger neighboring volcano, Mauna Loa, which erupted several times between 1912 and 1940. Logistics alone were daunting on this nearly 14,000-foot-tall mountain. Jaggar's initial ascents to the summit traversed the south flank by way of the 'Ainapō foot trail across the Kapāpala Ranch. During my stint at HVO, I traversed the lower part of this route easily by jeep, but in 1913, the trail crossed rough and spiny lava flows that made travel difficult even for mules. By 1916, spurred by Jaggar's tenacity and drive, the U.S. Army completed a reasonably smooth foot trail up the east flank of Mauna Loa, providing access for scientists and tourists alike. Accommodations at the summit, however, were less than comfortable. The only shelter was a gaping fissure partly covered by sheet metal, and water was available only from another nearby crack, where moisture accumulated.

The potential for lava flows to enter and damage towns lower on the slopes of Mauna Loa concerned Jaggar. He watched helplessly in 1926 as a lava flow on Mauna Loa's western flank destroyed the coastal village of Ho'ōpūloa. And he worried about Hilo, the island's largest city, which lay directly in the path of potential flows on the east flank of the volcano. With characteristic energy and creativity, Jaggar designed a system of barricades to divert such flows around Hilo, and during the eruption of 1935, he used explosives to reroute a lava channel and divert the flow from its Hilo-directed path. The system of barricades was never constructed, and his experiment with explosives was inconclusive. But scientists continue to apply both of those techniques with limited degrees of success at various volcanoes around the globe, including Mauna Loa and Kīlauea, as they try to control or at least modify the will of nature.

Jaggar also initiated drilling as a research tool on Kīlauea. With the type of primitive drill rig available during his HVO tenure, he could penetrate only about one hundred feet into the hard and unpredictably inhomogeneous volcanic rocks. Nonetheless, his pioneering efforts may have presaged extensive programs of research drilling, sometimes into pockets of magma, that came to HVO in the 1960s and 1970s and stepped out to other places, near Hilo for example, by the 1990s. When I joined the HVO staff in 1969, my principal assignment was to oversee ongoing drilling through cooled lava crust into a pool of magma that had recently spilled into Kīlauea's 'Alae Crater. However, through unfortunate timing—or perhaps at the will of the Hawaiian volcano goddess, Pele—renewed nearby eruption added more lava to the ʻAlae pool, which engulfed and totally buried the drill rig just days before my arrival. This proved to be only the first of many experiences that made me wonder whether we scientists, or some other powers, set limits to research at Kīlauea.

With typical, if accidental, foresight, Jaggar began the study of volcanic fumes, the volatile constituents of Kīlauea's magma. These most ethereal of all erupted products drive an eruption's vigor and explosivity. An erupting volcano acts much like a vigorously shaken can of carbonated beverage, which spews forth its contents when opened. Pressurized magma arriving at the earth's surface rapidly releases its volcanic gas, shooting towering fountains of lava skyward in aerosol-like sprays. The sprays can blanket terrain near and far with hot clots of molten lava and its solidified products, cinders and ash.

As vapor at the atmospheric pressure of the earth's surface, the volcanic gases tend to disperse and mix with the atmosphere before scientists can collect them for later study. However, Jaggar and his colleagues trapped gas samples escaping from the surface of the lava lake in Halemaʻumaʻu. The scientists held steel poles that extended glass tubes under vacuum just above the surface of the lava lake. When the researchers broke the end of the glass container against the lava crust, the vacuum sucked in nearly pure magmatic gas. The researchers subsequently demonstrated that such volcanic emanations contain mainly dihydrogen oxide (H_2O), which is common water when at room temperature and pressure. At 2,000 degrees Fahrenheit, however, water in erupting magma forms bubbles of super-hot steam that burst violently into the atmosphere. Jaggar and his team also found carbon dioxide and sulfurous gases in the samples.

Nearly forty years passed after Jaggar and coworkers sampled and analyzed Kīlauean gases at Halemaʻumaʻu before renewed investigations of lava gases gained momentum as improved sampling and analytical technologies became available. One might wonder why scientists had not focused more on the gases, since they drive the towering fountains of lava at Kīlauea. Samples of lava gases that Earth's atmosphere has not adulterated are commonly difficult, as well as dangerous, to collect. Moreover, though visually impressive, lava fountains at Kīlauea and similar basaltic volcanoes are not particularly powerful or threatening to humans and their material trappings, unless a fountain happens to spew out of one's backyard. That is an unlikely scenario for the half of Kīlauea that lies within

Scientists attempt to collect volcanic gases at Halema'uma'u lava lake. A glass tube under vacuum is attached to the end of a steel pole. Once positioned over the active lava, the scientist breaks the end of the tube by tapping it against the thin lava crust, prompting gas to be sucked in. Heat from the lava melts the broken end of the tube, resealing it. Note that scientists of the early 1900s tended to dress more formally for field work than those of the late 1900s.
—*Journal of Geology* photo

Hawaiian Volcanoes National Park, and for the rest of Kīlauea, including the lower part of the east rift zone, which is sparsely inhabited and rural. Thus, immediate threats of volcanic hazards did not fuel a need and desire to study the gases, and intellectual curiosity apparently lagged.

The study of volcanic gases has graduated from Jaggar's primitive experiments to become an important aspect of volcanology. Scientists continue to study how volcanic gases affect the violence of eruption, how they create valuable mineral deposits, and how they contribute to atmospheric pollution (especially the carbon dioxide and sulfurous gases of magma). Since the early 1980s, a nearly continuous outpouring of lava from the east rift zone of Kīlauea into the sea has created a downwind haze of volcanic gases mixed with partly vaporized ocean brine. This plume of volcanic smog, popularly known as "vog," can be harmful to the health of people with even minor respiratory ailments. While volcanologists cannot cure this ill, they can advise on what the pollutants are and when they are being produced.

Volcanologists even study how the changing release of gases through time from the surface of a volcano might help them predict an impending eruption, at Kīlauea and at volcanoes around the world. And it all began with a few glass containers under vacuum.

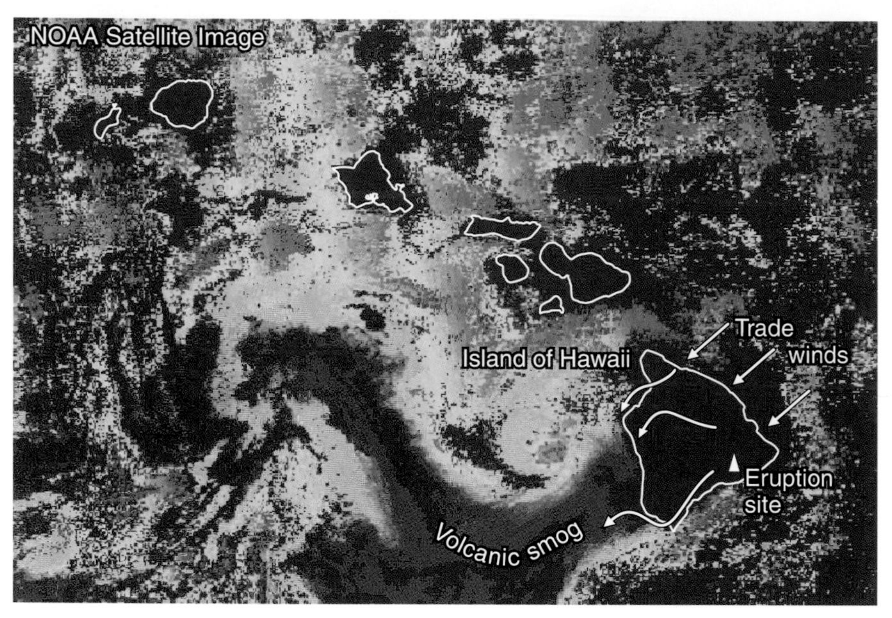

Volcanic smog, popularly called "vog," appears as a yellow-orange-red plume in this satellite image. Trade winds carry vog hundreds of miles downwind from its point of origin, where lava from the indicated eruption site flows downslope and spills into the Pacific Ocean.
—NOAA satellite image

By the time he retired from HVO in 1940 to continue research at the University of Hawai'i in Honolulu, Jaggar had created a legacy previously unparalleled in volcanology. He had plenty of help along the way, with such eminent scientist coworkers as A. L. Day, Reginald Daly, Ruy Finch, Frank Perret, and E. S. Shepherd; with local Hawai'i volcano enthusiasts such as Lorrin Thurston; and with the many manual laborers whose muscle and sweat made Jaggar's experiments a reality. But HVO was essentially Jaggar's creation. He provided scientific and spiritual continuity from the beginning until he retired from HVO, while most others came and went. In so doing, he set the table for a feast of increasingly sophisticated understanding of how Hawaiian volcanoes work—a feast that a growing staff of innovative scientists and their steadily improving technologies further flavored and savored during the next six decades and beyond.

The year Jaggar left HVO was also the year my parents created a little zygote that developed into me—and I took my first figurative squirm toward my HVO destiny.

Thomas Jaggar examines a huge boulder of basalt that blasted out of Halema'uma'u Crater in 1924 during a violent steam explosion, a phreatic eruption in geological parlance.
—Courtesy of Bishop Museum

3

THE HAWAIIAN VOLCANO OBSERVATORY GROWS

After Jaggar, the HVO leadership baton passed to Ruy Finch, to Gordon Macdonald, and then to Jerry Eaton, who added a staggeringly impressive list of scientific accomplishments to the history of the observatory. Eaton is a broadly trained earth scientist who specializes in seismology, the study of earthquakes. He transferred to HVO in 1953 from his U.S. Geological Survey post on the mainland. During seven remarkable years there, he wrote a landmark paper (with K. J. Murata) describing how volcanoes grow (see Additional Readings). Eaton's principal conclusions in that paper have withstood the scientific tests of time into the twenty-first century. Whereas Thomas Jaggar had broadly—and necessarily speculatively—outlined Kīlauea's behavior, at least for the summit area, Jerry Eaton filled in a multitude of details that tested and fundamentally confirmed Jaggar's model as scientifically sound. By 1960, a finely blended mixture of their scientific accomplishments produced a Tom-and-Jerry elixir that was smooth nectar, palatable to all tasters.

Eaton spent most of the years 1953 through 1961 at HVO, five of them as Scientist-in-Charge, as the director's position had come to be called. Like Jaggar, he understood the fundamental importance of seismology to charting volcano behavior, but unlike Jaggar, he had at his disposal enhanced resources and improved technologies. By 1958, Eaton had installed four seismometers around the summit area of Kīlauea, all linked by hand-laid wire to a common clock. The task of emplacing miles and miles of this wire was tedious and physically exhausting, but such coordinated timing in a network of seismometers is absolutely necessary to accurately determine where earthquakes originate within the earth. With accurate data in hand, Eaton soon showed that quakes within and directly beneath Kīlauea originate at depths from almost at the surface to about 35 miles beneath the surface, where mantle-derived magma apparently enters a system of conduits that funnel it upward into the volcano. Moreover, this charting of earthquakes in effect closely defined the subterranean passageways magma traveled en route to sites of eruption.

While Eaton was modernizing the HVO seismic network, I was just becoming aware of geology, both the word and the career possibilities. My epiphany came accidentally by way of a science teacher during my junior year in high school. Largely departing from the subject at hand, this teacher spent much of one lecture speaking wistfully of his brother-in-law's career as a geologist. In so doing, he led the class's imagination on a romantic adventure in search of oil fields in the Middle East. The oil patch was not to be in my future, but the allure of a geologist's active, outdoor life stayed with me as I pursued my education.

Meanwhile, back at HVO, Eaton was interpreting an unusual type of volcanic earth shaking. Called harmonic tremor for its steady shaking of constant intensity, this type of earthquake lasts from minutes to as long as hours and lacks an abrupt beginning and end. He recognized that harmonic tremors were associated in time and space both with eruptions and with great swarms of "conventional" volcanically triggered earthquakes. He concluded that steady earth shaking was evidence that magma was streaming and pulsing through underground conduits. This conclusion stands firm today, in embellished form, and is a first-order basis for forecasting impending eruptions at volcanoes worldwide.

Though he continued Jaggar's work of measuring and interpreting ground tilt indicated by "out-of-plumb" seismometers, Eaton understood that he needed a truly portable and highly sensitive tiltmeter if he hoped to establish a geographically widespread pattern of tilt rather than document what was happening at a single point. So he designed just such an instrument. The HVO machine shop fabricated the instrument from U.S. military trash—the casings of large-caliber artillery shells, which were readily available on the Big Island.

Field tests quickly proved that the new tiltmeter could detect an almost imperceptibly small change in the inclination of the earth's surface—the change equivalent to raising or lowering one end of a 78-mile-long surface a measly 1 inch. Then Eaton established multiple tilt stations strategically located to circumscribe the summit area. The summit area soon proved to be the unequivocal focus of uplift and subsidence (equivalent to outward and inward tilt, respectively) related to the inflation and deflation of an underlying reservoir of magma.

Eaton's tiltmeter was still in use when I arrived at HVO in 1969. I had the late-night experience of using that tool to accurately document the ups and downs of the magma-gorged "balloon" beneath the summit of Kīlauea. We worked at night to minimize the disturbing effects of thermal expansion and contraction that come with the fickle sunshine of daylight hours. About the time Eaton was designing, building, and field-testing his portable tiltmeter, I was entering my undergraduate years at Carleton College, in Northfield, Minnesota. My declared major was geology, another step toward a volcano-rich destiny. Little did that Carleton geology student know he would one day carry out research by lying on cold muddy ground to read the finely graduated scale of Eaton's tiltmeter.

Well before Eaton left HVO in 1961, he increased the seismic network to about a dozen stations, expanding coverage outward from the summit area to learn more about how the outer slopes of the volcano behave.

From HVO's beginnings in 1912 up to the 1955 eruption along the volcano's east rift zone, research at Kīlauea tended to focus on the summit caldera and its immediate surroundings, if only because that area was easily accessible and visually captivating. Besides, a nearly nonstop extravaganza of eruptions played in Halemaʻumaʻu during the first three decades of HVO's existence. That was where the action was. But many eruptions at Kīlauea occur on the flanks, along zones of cracks and rifts that radiate from the summit caldera, and knowledge of volcanism within these zones would help scientists better understand the complete volcano.

Kīlauea's flanks slope mainly east and south, into the sea. Eaton simply *had* to direct some of his efforts at Kīlauea's east rift zone, which erupted in

1955 and again in 1960. He seized these events as opportunities to chart a well-integrated system of magma plumbing. Mapping of earthquakes and ground tilt defined a source region for magma in the mantle, which feeds into a magma reservoir within the volcano itself. Subsidiary pathways carry magma from this reservoir to eruption sites within the summit caldera and along rift zones. With this information, Eaton established a model of the volcano. According to Eaton's model, a magma reservoir beneath the caldera episodically inflates and deflates like a balloon. These episodes corresponded with doming and subsidence on the volcano's surface. Magma leaks to the summit and rift zones produced eruptions that punctuated the times of uplift and subsidence.

Jerry Eaton moved back to the mainland in 1961 to pursue his seismic research there. But the work he had begun, including studies of eruptions on the flanks of Kīlauea, continued in the hands of other geologists.

Even if flank eruptions had been common during the early years of HVO, the lack of an instrument to measure multimile distances with accuracy, rapidity, and relative ease would have stifled effective study of ground deformation. This impediment disappeared in the mid-1960s, when a portable machine known as an electronic-distance-measuring device, or EDM, became available.

EDM surveys could cover more ground faster and define more accurately ever-smaller changes in the horizontal position of a surveying reference mark—thus precisely quantifying horizontal ground deformation, such as doming and subsidence. Almost overnight, EDM surveys replaced the clumsy, time-consuming, and imprecise technique of triangulation. In place of triangulation, the new surveying method used trilateration, measuring the lengths of the legs of a triangle rather than the internal angles. Scientists had already been using the most accurate technique of leveling to determine elevation, but the introduction of the EDM allowed the HVO team to study horizontal ground deformation on the volcano's flanks.

First came a shakedown cruise, using an EDM to document horizontal ground movements in the summit area. Confidence in the EDM grew rapidly as initial results proved to be entirely consistent with those of leveling and tilting. Shortly after I joined the HVO staff in 1969, our team was in full-EDM flight across the entire surface of Kīlauea. We began an extensive trilateration survey where only triangulation had gone before. We compared our new trilateration data with the oldest triangulation survey for the area and learned that parts of Kīlauea had moved seaward as much as

With the broad profile of Mauna Loa Volcano as a backdrop, HVO staff measure the distance between the EDM device, whose red laser beam is visible here, and a reflector positioned elsewhere on Kīlauea Volcano. —HVO staff photo, U.S. Geological Survey

14 feet since the beginning of the twentieth century. This movement was an integral part of the volcano's growth. Thus, we could now view the ups and downs of the summit area, associated with the inflation and deflation of the summit magma reservoir, as a fairly localized and schizophrenic bulge on a much-broader, volcano-wide spreading.

By the early 1990s, satellite-based surveys, which measure positions of reference marks on Earth, had simplified data gathering. The satellite-based instrumentation, known as a global positioning system, or GPS, also provides a frame of reference not available to EDM surveys. To define horizontal movement of ground, EDM surveys had to assume that some part of the surveyed area of the volcano stayed motionless while other parts moved relative to that "stable" area. By contrast, the GPS frame of reference is the satellite itself, which requires no assumption about which parts of the volcano may be mobile. The GPS surveys comprehensively documented the volcano's mobility, verifying most conclusions drawn earlier from EDM and triangulation surveys. GPS is in the process of replacing the older surveying techniques for volcano-wide studies. The most pertinent and now-answerable question is no longer "does the whole volcano deform as it grows?" but rather "at what rate and in what changing patterns does it deform?" Such knowledge is a powerful tool in ongoing attempts to forecast eruptions and related volcanic hazards.

The related volcanic hazard of foremost interest is a landslide, or series of landslides, of potentially gigantic proportions. These occur once the seaward-facing flank of the volcano has bulged so far that it no longer has enough cohesive strength to withstand the pull of gravity. The headwalls of such landslides, some hundreds of feet tall, scallop much of the seaward-facing slopes of Kīlauea. The EDM and, later, GPS data help the HVO staff understand why the volcano is so broken and disrupted—scalloped by landslide headwalls and cracked and fractured along rift zones. Almost simultaneously with the introduction of GPS surveys, new offshore charts of the configuration of the seafloor detailed the jumbled and tumbled look of the surfaces of huge landslides and debris avalanches. Not surprisingly, the distal ends of Kīlauea's landslides extend well below sea level, hidden by deep Pacific waters.

Also hidden on the ocean floor is evidence that virtually all of Kīlauea—not just landslide blocks from high up in the volcano itself—moves seaward. This "hidden" evidence came into view when a very energetic earthquake of magnitude 7.2 struck at the base of the volcano in 1975. By that time, the network of seismometers was volcano wide. Techniques to analyze earth shaking were so advanced that scientists could determine that this particular quake represented jerky seaward movement along a nearly horizontal plane of slippage at the base of the volcano. This marks the depth at which volcanic rock rests on a thick layer of loose and slippery seafloor mud, hardly a solid foundation for Pele's home. So, whereas the EDM and GPS data show that the surface of Kīlauea moves seaward, this earthquake information showed that the very bottom of the volcano moves as well. It seems the entire volcano, from top to bottom and side to side, moves toward the sea.

Many other topical studies have contributed to an understanding of Kīlauea Volcano during the post-Jaggar years. For example, geologic mapping that began in the 1960s provided information needed to help quantify

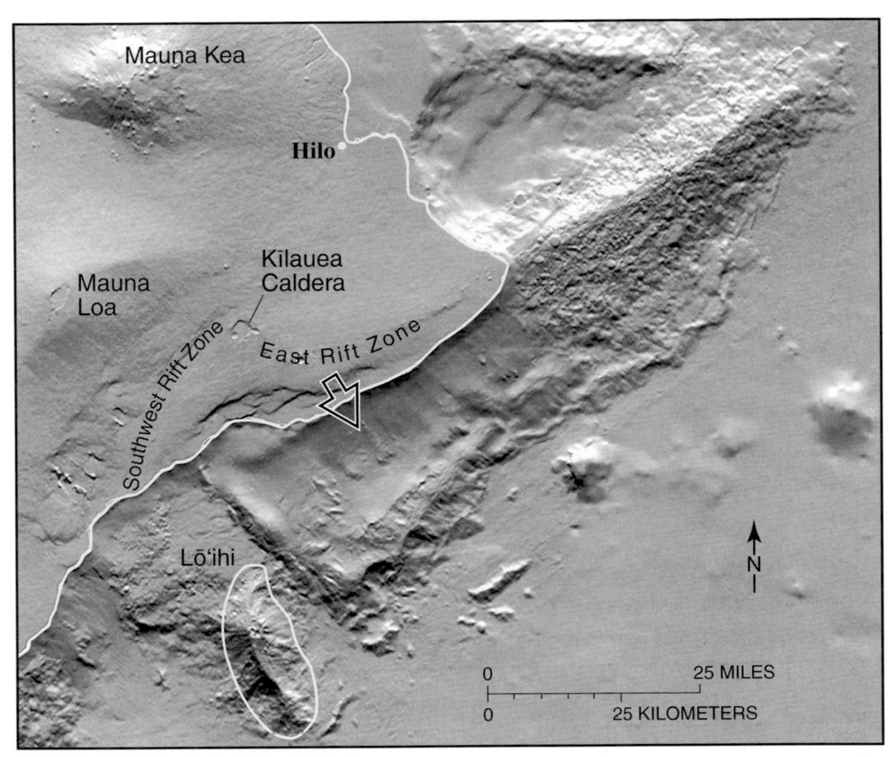

This digital elevation model shows the shape of Kīlauea both above and below sea level. Risers of the grand staircase cast shadows across much of the above-sea-level south flank of Kīlauea. These are the headwalls of huge landslides, which, together with southward movement of the entire volcano along its seafloor base, produce the smooth to jumbled and hummocky submarine landscape northeast of Lō'ihi. The large arrow shows the approximate horizontal direction of movement as documented by measurements made on land.
—DEM created by William Chadwick

where and at what rate new lavas were added to the mountain. Subsequent volcano-wide mapping suggests that lavas no older than about 1,000 years cover 90 percent of the present surface of Kīlauea. That is an alarming rate of resurfacing for terrain that people have inhabited, albeit sparsely, for only about 1,600 years.

Researchers have examined the chemical and mineral compositions of Kīlauea's lavas over the years, especially those of historic age. All these dark rocks are a type of basalt that geologists call tholeiite, and their compositions vary only slightly from one another. Still, the subtle variations in composition generally tell scientists something about the system of subterranean plumbing passageways and magma-storage chambers, which

The original HVO building was modest in size and appearance. —Courtesy of Bishop Museum

By 1958, HVO operations were housed in substantially larger quarters, across the caldera from the original location. —U.S. Geological Survey photo

By 1987, HVO had expanded into an entirely new and larger building, in addition to the nearer part of the earlier L-shaped space. The far leg of the L is today's Jaggar Museum. Halema'uma'u is in the background. —U.S. Geological Survey photo

improved seismic and ground-deformation studies have described in ever-increasing detail. The overall story that relates magma plumbing to lava compositions is remarkably coherent.

The physical plant of HVO has evolved with changing needs for space and equipment. In 1948, operations moved from the Whitney Laboratory of Seismology to a modest yet larger building at Uwēkahuna Bluff. In 1958, even this "new" facility had to be nearly doubled in size to accommodate a growing staff with increasingly more ambitious research projects. Within another twenty years, HVO outgrew this enlarged laboratory, and by 1987 an entirely new building was completed adjacent to the smaller Uwēkahuna structures. The initial, lone, small building there reverted to its original use as a museum for Park visitors. It is now named the Jaggar Museum, in honor of HVO's founder.

As HVO enters the twenty-first century in its comfortably spacious quarters, with ninety years of experience and results under its belt, current knowledge of how Kīlauea Volcano behaves continues to accumulate mainly by extending lines of research begun during the Jaggar years. Jaggar established programs to monitor ground deformation, seismicity, properties of magma, and even those ethereal gases so difficult to sample. Time and new technologies have increased our knowledge of volcanic behavior and will continue to do so. Still, Jaggar's vision in founding the volcano observatory laid out avenues of study that geologists have faithfully and fruitfully followed over the ensuing years.

4

MINGO LEADS THE WAY

In 1969, Neil Armstrong took his "one giant leap for mankind" as he set foot on the Moon. And with no wind or flowing water on it to erase them, his space-age tread marks are probably as fresh today as on the day Armstrong made them. While Armstrong made history in space, my wife, Anne, and I were planning our first major earthly journey together. Our adventure would take us only halfway across the Pacific Ocean to Hawai'i, yet it was an Armstrong-like discovery trip for us, young newlyweds who had never before tasted life in the tropics. I had recently completed my Ph.D. in geology at Stanford University and had begun a career with the U.S. Geological Survey. Shortly thereafter, as luck would have it, the U.S. Geological Survey asked me to serve on the staff of the Hawaiian Volcano Observatory, starting in August 1969.

And so it was that Anne and I looked up at that human-carrying July 20 moon from the yard of our tract bungalow in Palo Alto, California, knowing we would soon be traveling to Hawai'i. Our cat, Mingo, joined us in our moon-watch vigil.

In 1969, Mingo—short for Domingo, the name of a special friend from Mexico—was our one-and-only pet. He served the dual functions of surrogate child and client for a frustrated veterinarian, namely Anne. As an about-to-graduate high-school student, Anne had planned for a career in veterinary medicine, but she was a woman born of the wrong generation. Adults humored her dreams, then advised her to attend a nice liberal arts college and earn her Mrs. degree. She did indeed attend a nice liberal arts college, Carleton College in Northfield, Minnesota. So did I, and while there, I helped her secure the Mrs. degree. Still, the idea of being a vet never really left Anne's mind, and Mingo was just one in what was to be a veritable Noah's Ark of animals that would adorn and enhance our lives during the coming years.

A cat of unknown and questionable parentage, Mingo had a pure black coat, and his eyes were a deep emerald green. We found him in 1966 as a barely weaned kitten at an animal shelter. He became ours for one dollar and a promise that we would have him neutered before he could father litters of alley cats. We held to this promise and added surgery to remove front claws when Mingo showed a fondness for shredding the upholstery.

Mingo was firmly in our Hawaiian plans. At two years old, he had become as much a part of our lives as the Pontiac or TV. As a living creature in a household that valued such pets, Mingo was far less easily replaceable than material things. We soon discovered, though, that sending a cat, or any other animal, to Hawai'i is complicated for a reason that transcends bureaucratic nonsense.

Hawai'i is one of those few remaining places on Earth that is rabies free. Having emerged from the ocean floor as pristine, isolated, and disease-free volcanoes, Hawai'i has so far been spared the introduction of rabies, despite centuries of human and other animal visitations from many infected parts of the world. Whether this rabies-free condition is the product of luck or good planning, it is a fact. So Mingo would be granted permission to enter the state only after a 120-day quarantine.

We understood and respected the reason for the extended quarantine and planned to abide by the rules. But, as lovable and special as Mingo was to us, he did not fit the organizational mold of my employer. Mingo very nearly caused the agency to cancel my invitation to Hawai'i when word got out that we planned to take this household pet with us.

Now, we raised Mingo to be a house cat who specialized in an unending repetition of eating, sleeping, and using the sandbox, a level of inactivity much like that of most other house cats we've known. By moving time in late 1969, he was thus a pretty tractable and harmless guy. Even though his tell-tale black hairs often made their way into my homemade sandwiches and onto all light-colored objects in our house, he hardly seemed a valid reason to deny me the opportunity to study Hawaiian volcanoes.

The problem was that the U.S. Geological Survey, like many long-lived and partly ossified bureaucracies, sometimes seems to live by the motto "We can't do it because we haven't done it before." And apparently no U.S. Geological Survey geologist before me had ever asked to move a cat to Hawai'i along with the traditional spouse, furniture, and car. When I announced that a cat would accompany me to Hawai'i as part of the household goods, great hand-wringing and consternation ensued, beginning with my supervisor and escalating toward the apex of the organizational power pyramid.

I soon learned that the problem was even more serious in bureaucratic minds than just upsetting the regular protocol of the U.S. Geological Survey. In their efforts to dissuade me from taking Mingo with me, the bosses also explained that each U.S. Geological Survey employee who was temporarily transferred from the mainland to HVO was expected to live in a house provided by the National Park Service. A long-standing policy at Hawai'i Volcanoes National Park, the home of Kīlauea, strictly prohibited the keeping of household pets within the park. I was asked to believe that chaos would befall this tidy arrangement if even just one U.S. Geological Survey short-termer were to upend the status quo. I viewed my request as completely reasonable, but the supervisor who made the final decision thought I was more than a peck of trouble.

Following many weeks of discussion and negotiation, I was granted permission to take Mingo and live outside the National Park. What had then seemed stupid to some was viewed as enlightened policy soon thereafter—a pacesetter in many ways, Mingo had established a beachhead. About a year later, with the benefit of hindsight, all interested parties agreed that Mingo led the way to beneficial changes in the system.

Permission finally in hand, making arrangements for Mingo's actual travel to Hawai'i was remarkably simple. For a price, the airlines were happy to ship Mingo to the Islands and to the quarantine facility. They guaranteed a comfortable flight and assured that our beloved cat would be delivered safely to the 120-day holding facility. So, with only a little trepidation we loaded Mingo into a sturdy traveling kennel, took him to catch a nonstop flight to Hawai'i, and bid him a temporary adieu as his jumbo jet headed west. The next day, we were notified that he was safely resting in his Hawaiian quarantine quarters. Friends living in Hawai'i agreed to visit him periodically and let

us know how he was faring. Their updates described an extremely comfortable and self-satisfied animal. Apparently, even life in quarantine is enjoyable in the tropics.

Our dream of walking on new lava flows and sandy Hawaiian beaches about to come true, Anne and I just needed to get ourselves moved.

5

OUR NEW HOME

As anyone who has experienced it knows, moving a household is no picnic. And the process becomes less fun to repeat as we accumulate more and more stuff. But our move to Kīlauea was easy and low stress. First, we were so young and recently married and my career was so new that we owned few things. Then, as if the hand of providence was reaching to help, our house was burglarized twice during the two months before our move, reducing the few material things that needed shipping. We were selling the house, and professional movers would pack and ship the household goods that apparently were not up to our burglar's standards. The Pontiac would be shipped from a nearby drop-off station. That left only the clothes on our backs and in our hand luggage for us to handle.

And, of course, we just plain wanted to move. Who would not want to move to Hawai'i, especially when paid to do so? The Pontiac and household goods disappeared in late August 1969. A month after Neil Armstrong's walk on the moon, we boarded the flight that would launch us on a giant leap in our young lives.

With Mingo at about forty days in quarantine and counting down from the 120-day mark, we made our temporary quarters in one of the houses in Hawai'i Volcanoes National Park normally reserved for U.S. Geological Survey employees. From that base, we searched for and quickly found a house to rent about a mile outside the park, in a village appropriately named Volcano.

Home for the next three years was to be an unimaginative 700-square-foot rectangle, perched about 4 feet off the ground on several inadequate-looking wooden posts. We soon learned that these spindly posts, though weak in appearance, strongly amplified the steady shaking of the ground that typically accompanied the volcano's eruptions.

Though not very pleasing to the eye, inside or out, this house seemed entirely adequate for our needs. A bedroom and a living room of roughly

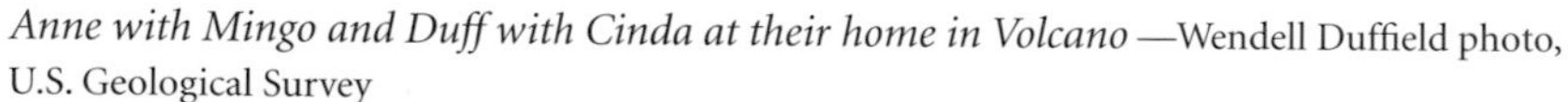

Anne with Mingo and Duff with Cinda at their home in Volcano —Wendell Duffield photo, U.S. Geological Survey

equal sizes dominated the floor plan, complemented by a small kitchen and a tiny bathroom. Two internal walls subdivided the overall house rectangle into these four rectangular spaces. Pure simplicity and functionality. Though it was the smallest room, the bathroom proved to be vast in pleasure giving. For after many a long day of chasing lava through a rainy forest, I took sublime refuge in the room's principal fixture: an antique, porcelain-covered steel tub perched on four majestic clawed feet.

The outer walls of our boxy house were 1-inch-thick boards, exposed both inside and outside the structure. While shopping for a house to rent, we had noticed that this was a widely practiced construction technique in Hawai'i. Though foreign to our thinking, so seasoned in severe winters were we, such single-board construction stems from the logic that homes in the tropics need not be built to withstand winter chills. Such logic, however, neglects the fact that elevation as well as latitude controls the diurnal temperature range, whatever the season. So, tropical latitude notwithstanding, the nearly 4,000-foot elevation of Volcano quickly drove us to a hardware store to purchase a portable kerosene heater, which we carried from room to room to soften the typical evening chill. Thank goodness for having so few rooms.

With none of the dead, and normally insulated, air space typical of double-wall construction, exposed water pipes and electrical conduits decorated both inside and outside surfaces of our Hawaiian place. We suspended our own wall decorations from holders whose nails penetrated wallboards, their sharp ends protruding outside. With the house up so high on posts, those sharp ends of picture hangars could not impale visitors who might happen to rub against an exterior wall.

The roof was corrugated metal that noisily reminded us of its presence with every rain—and with an average rainfall of about 160 inches per year, the reminders were frequent. Our household water came from a large wooden storage tank perched above ground adjacent to a corner of the house. The tank sat at a level just low enough for rain gutters to feed in the abundant runoff from the roof. We never once ran short of water in our three years there. Typically, the tank overflowed day and night, splashing its surplus on a flat rock, which added nearly constant background noise over which the cacophony of rainfall played.

Landscaping around the house was the product of an endless battle with natural vegetation that thrived on the abundant rainfall. Our house, like all our neighbors' houses, stood in an artificial clearing in the rain forest, and many of our neighbors labored tirelessly to maintain landscape plantings of their choice. Our idea of landscaping the clearing that held our house was to keep the perimeter at about the same position so the house would not incrementally disappear into a natural thicket of trees, ferns, and grass.

We had another special reason to maintain the clearing around our house. During the few, brief moments when the clouds receded, the view from our house was extraordinary. The house's one picture window framed the broad and gentle slopes of Mauna Loa Volcano, Kīlauea's "big sister," on the horizon to the northwest. From an appropriate position in the living room, we could see the rounded, nearly 14,000-foot-high summit of Mauna Loa. With the changes in seasons, the peak changed from lusterless black to a shimmering white cap, a reminder that snow can fall even in the tropics if the landscape rises high enough. Not a bad view from a small and ugly box perched on shaky stilts in the rain forest of Hawaiʻi.

Several hundred people, most permanent residents, populated our adopted village of Volcano. Thick and tall vegetation of the rain forest provided an effective visual screen that completely hid most structures from

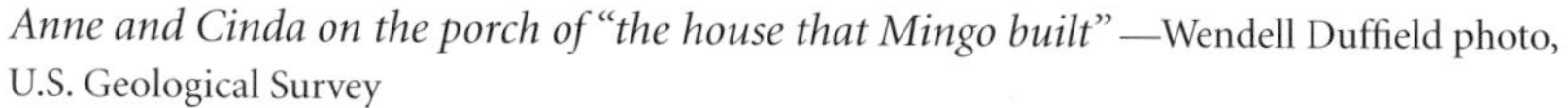

Anne and Cinda on the porch of "the house that Mingo built" —Wendell Duffield photo, U.S. Geological Survey

the view of a passing motorist. The vegetation created a degree of privacy intruded upon only when one left the isolation of home. Though we were outsiders in terms of our birth origins, this "hidden" community warmly welcomed us. We were quickly adopted into a spirited bridge club and soon became recognized regulars at the only local commercial enterprises, Okamura (now Volcano) Store and Hongo (now Kīlauea General) Store.

These were general stores of the old school—shelves chaotically stocked with a little of everything and a lot of nothing—though modern enough to offer fuel for automobiles. Serious shopping needed the big-city facilities of nearby Hilo. However, the supplies from the local stores could keep a household functioning at the pace of life in Volcano for weeks on end. Okamura Store offered a marginally wider variety of goods, but Hongo Store served as the local post office, and so got about as many customers as the competition a couple of hundred yards up the road. The stores' most valuable function was not to provide household supplies but to promote social intercourse.

By the time our furniture and Pontiac arrived from the mainland, about one hundred days into Mingo's quarantine, we could see that life in our tropical community was going to be serene and comfortable, as well as wet, humid, and sometimes moldy. Our rental house in Volcano, which we thought of as "the house that Mingo built," would be a fine base from which to explore the enticements of Hawai'i.

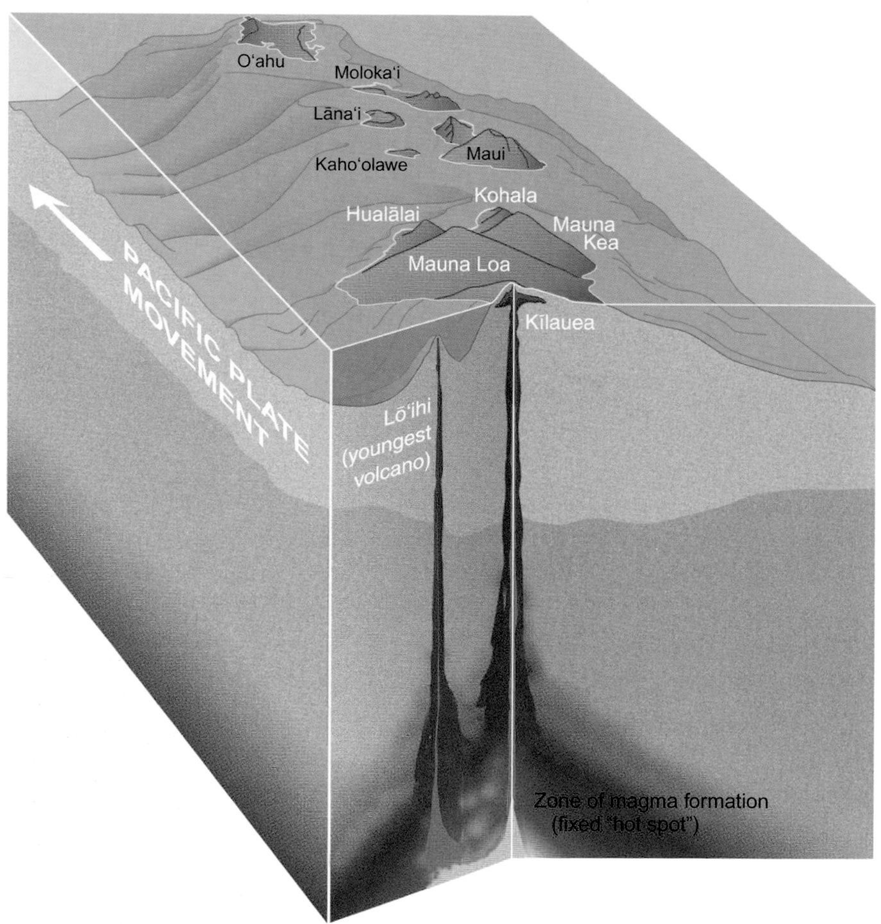

As the Pacific plate moves over the stationary underlying mantle, magma rises through the plate and erupts to form a chain of volcanoes on the seafloor.

6 FOCUSING IN: THE HAWAIIAN ISLANDS AND THEIR STRING OF OLD NEIGHBORS

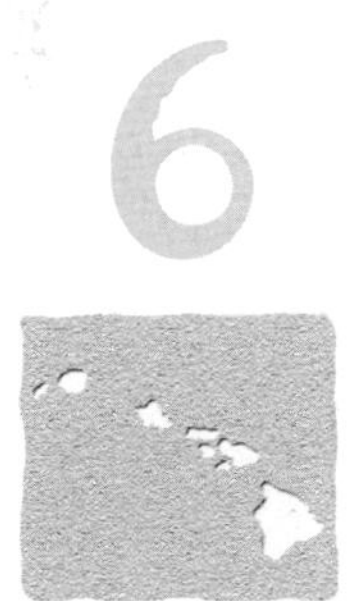

A cursory glance at a map shows the Hawaiian Islands stretched like a string of pearls across the mid-Pacific Ocean. About a century ago, Mark Twain described these as "the loveliest fleet of islands that lies anchored in any ocean." At about the same time, the American geologist James Dwight Dana noted that not all ships of this fleet appear equal. Dana saw that the islands become progressively more eroded, and therefore older looking, from the southeast to the northwest along the island chain.

Both of these observers were describing what scientists call an age-progressive chain of volcanoes. The islands in the chain have formed sequentially as the floor of the Pacific Ocean has moved slowly and steadily over a stationary and persistent source of magma, called a hot spot, in the mantle.

The Hawaiian Islands are part of a very long volcanic mountain range in the sea, with some of the peaks in the range standing above sea level and some below. The oldest volcano in the chain, called Meiji, began to rise from the seafloor about 80 million years ago. The newest volcano, Lōʻihi, is growing today and will probably continue to build for a few thousand years. If we could drain the Pacific Ocean and then look down from space, the Hawaiian Islands would appear as the most southeastern, and youngest, end of the chain. The older end is nearly 3,700 miles away, in a remote corner of the Pacific where the Aleutian Islands of westernmost Alaska meet the Kamchatka Peninsula of Russia. In the view from space, two features of this chain would catch our attention.

First, even the casual viewer would notice that the chain is not laid out in a line. Instead, a kink separates the mountain range into two linear segments of roughly equal lengths. The two linear segments represent nearly equal durations of time, the bend occurring at a volcano that formed about 40 million years ago.

The observant viewer would also notice that the volcanic mountains are progressively smaller and look less like volcanoes west and north along the chain. The tallest and least eroded of the mountains lie within the southeastern-most part of the chain, in the Hawaiian Islands. Northwest of the Hawaiian Islands, even before the turn at the elbow, some of the mountains show no vestige of a peak and instead are nearly flat topped or even capped with a coral reef. Most of the mountains this far up the chain and beyond are called seamounts. They have lost their volcanic appearance through erosional decapitation. Very few of these older volcanic mountains even rise above sea level. That is why we can most readily establish the big picture after figuratively draining the Pacific Ocean.

Any conversation beyond casual reference to the Hawaiian Islands almost demands discussion of the name Hawaiʻi. This name carries more than one meaning, which can, and often does, result in unintended confusion.

The name Hawaiʻi refers to the group of islands—Niʻihau, Kauaʻi, Oʻahu, Molokaʻi, Lānaʻi, Maui, Kahoʻolawe, and Hawaiʻi—strung out from northwest to southeast in a 360-mile-long line. These are the islands that intrigued Twain and Dana. The same word also refers to the most southeasterly member of the group. By popular custom, the island Hawaiʻi is also called the Big Island, which conveniently circumvents ambiguity. Locals apparently know this duality of title from the moment of birth, whereas tourists seem to require a day or two in paradise to learn about it.

Then there is the political context of the name. The United States annexed Hawaiʻi in 1898, and until 1959, the U.S. government parlance for this group of

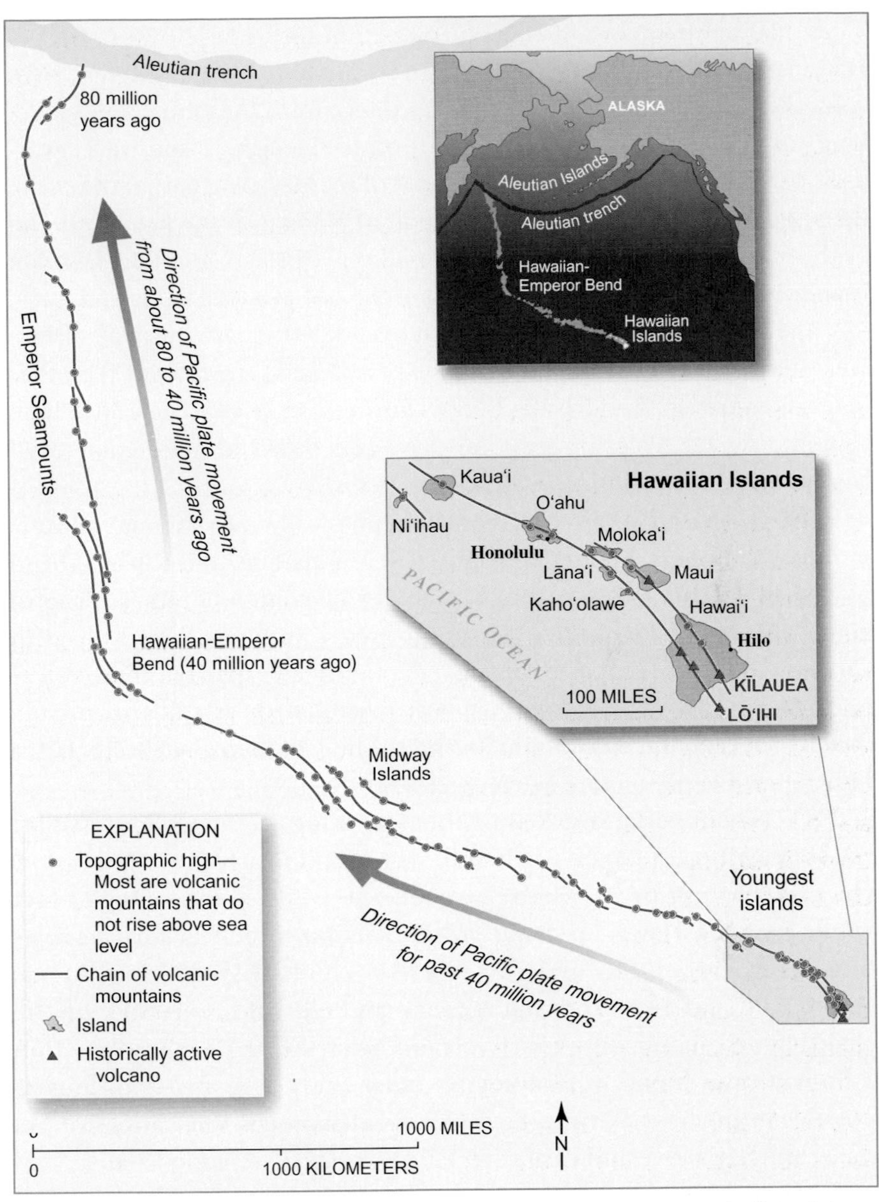

The setting of the Hawaiian Islands and their predecessors strung out across the floor of the Pacific Ocean

islands was the Trust Territory of Hawai'i. In 1959, the string of islands became the State of Hawai'i, the fiftieth member of the United States of America. And just to confuse matters further, the Big Island is the County of Hawai'i, as a state-government entity. What's in a politically generated name anyway?

In the more important context of earth history, Mark Twain's fleet of Hawaiian Islands fits comfortably into an 80-million-year age progression of volcanism. As one focuses in ever more closely, from the entire 3,700-mile-long volcanic chain that Hawai'i is part of, to the eight islands of Hawai'i itself, to the Big Island at the southeast end of the Hawaiian part of the chain, the age progression still obtains. The bulk of Ni'ihau, at the northwest end of the Hawaiian eight, grew about 5.5 million years ago, and the Hawaiian Islands become younger southeastward to the still-growing Big Island.

The Big Island moniker fits. At just over 4,000 square miles of surface area, the island of Hawai'i is almost twice as large as the total of the other magnificent seven. And the Big Island continues to grow as new lava flows spill into the sea. Volcanic eruptions have added several hundred acres of land to the island's south coast since the 1980s.

The Big Island itself consists of five physically overlapping volcanic mountains: Kohala, Mauna Kea, Hualālai, Mauna Loa, and Kīlauea, from northwest to southeast. Within the ability of contemporary science to define the age of each, these are progressively younger from Kohala to Kīlauea. Scientists recognize some overlap in the life spans of these volcanoes. After all, a volcano doesn't appear instantaneously full grown. Still, the birth of each volcanic mountain and the bulk of its growth pretty faithfully follow the big-picture age progression.

Yet another volcano, called Lō'ihi, is growing on the seafloor just off the southeast coast of the Big Island, its top still nearly 3,000 feet below sea level. Judging by its relatively small size, Lō'ihi is the youngest link in the growing Hawaiian chain. While it is not practical for scientists to focus studies on this undersea volcanic "child," they do learn lessons from Lō'ihi and even from the wizened and old volcanoes back up the chain. But to successfully chart volcano behavior and evolution within a human time frame, volcanologists must study as many eruptions as possible in the briefest time. So HVO scientists focus their attention on the eruptively active and easily accessible "teenage" volcano Kīlauea, and to a lesser extent the neighboring and much larger "adult" Mauna Loa. Kīlauea has erupted about sixty-five times since European visitors first arrived in the early 1800s.

The summit of Kīlauea Volcano stands about 4,000 feet above sea level. The village of Volcano is about 3 miles east of the summit, barely below this high point. To two kids who had grown up where any place above 1,000 feet in elevation was pretty high living, Kīlauea offered life in

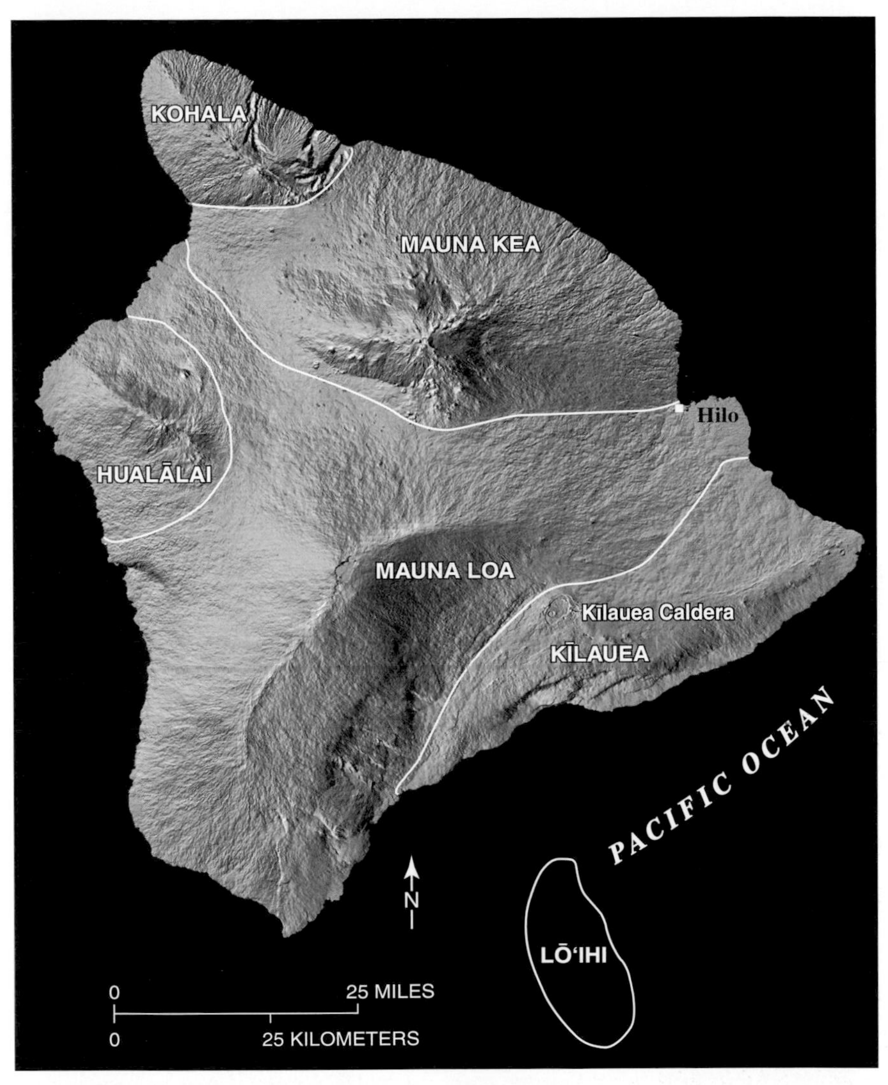

Digital elevation model of the Big Island, showing the surface boundaries between the volcanoes —Model created by Bob Mark, U.S. Geological Survey

the clouds. And with about 160 inches of rainfall each year, much of life in Volcano is lived in clouds.

The prevailing trade winds buffet the windward, or east, side of Kīlauea and account for the abundant rainfall and lush vegetation there. However, just a mile or two westward across the top of Kīlauea, persistent trade-wind weather, combined with the mountain's topography, create a rain shadow that receives only about one-fourth the precipitation typical on the mountain's wet side. Bare rock rules the landscape west of the summit, and the west flank of Kīlauea is called the Kaʻū Desert. Much of the west flank's barrenness, though, is likely due not to insufficient rain but to the hostile acidic growing environment created by rain that falls through the persistent volcanic fume blowing in from the summit. Our view of Kīlauea from outer space reveals a two-faced appearance of lush hairy green on the east and stark rocky gray on the west side of the volcano.

A basin roughly 400 feet deep and 2 miles wide indents the very top of the mountain. Geologists call this type of basin a caldera. The wall of the caldera is nearly vertical, as though newly shaped, and the rim gradually decreases in height toward the southwest. In 1971, lava flows overtopped the southwest rim and spilled out of the caldera basin.

From an overhead perspective, the almost circular shape of the caldera suggests a large cookie cutter removed part of the mountain and left the caldera's hole. Geologists will tell you, though, that the apparently missing part is still present—it just sank into the volcano. When molten rock, or magma, erupted from a magma reservoir not far beneath Kīlauea's summit and flowed away, the rigid top of the mountain lost its underpinning and collapsed. The volcano's surface collapsed onto the remaining magma in the underground reservoir. The caldera is the collapse basin.

The caldera we see today is not the simple product of one eruption and associated subsidence. Instead, the shape and size of the caldera is constantly changing, at least on the short-term geological time scale. Over the entire life cycle of a volcano like Kīlauea, calderas come and go. They form, fill with lava, reform perhaps in a slightly different location, fill with lava again, and so forth, reflecting the ebb and flow of magma into and partly through the mountain. Even over a human lifespan, the diameter and depth of Kīlauea Caldera have changed. Graphic written accounts and sketches document such change, starting in 1823 with those by missionaries, the first European visitors to Kīlauea.

An indentation left by yet another apparently missing piece marks the floor of today's caldera. A hole roughly 0.5 mile wide and 200 feet deep, Halemaʻumaʻu looks like a miniature version of the larger basin. All the summit eruptions take place through Halemaʻumaʻu and the larger caldera.

If Kīlauea's summit is the head of a squatting, south-facing volcano being, two tapering ridges represent the creature's outstretched arms. One arm extends eastward and the other southwestward, into the Pacific Ocean. Geologists call these limbs rift zones because they contain many fissures

This oblique aerial photo, taken in 1954, looks northward across Kīlauea Caldera. The circular crater within the caldera is Halema'uma'u, and the smaller oval crater just outside the caldera is Keanakāko'i. The top of Mauna Kea Volcano is in the left background.
—U.S. Geological Survey photo

and cracks from which rising magma spills. The above-sea-level part of the east rift zone is 35 miles long and tapers gently from the summit to the sea. In contrast, the western rift takes a much shorter and, therefore, steeper path to water. Kīlauea apparently has frequently exercised, by way of eruptions, and thereby built up the better-developed eastern arm more than the weaker-looking western limb. Still, each rift zone carries the potential to spew molten rock with the same dire consequences for anything in a lava flow's path. And as any victim would readily agree, burial by just one eruption is one too many.

The southward-facing slope of Kīlauea, between the two rift zones, is unimaginatively called the south flank. This terrain is passive in that it does not serve as the site of eruptions. That hot delight seems exclusively reserved for the summit and the rift zones. The south flank offers its own volcanic action, though, as the site of great earthquakes and the kind of earth breakage and scarring that results from such violent shaking. From time to time, earthquakes shake loose both small and huge pieces of the volcano's south flank and send them sliding toward the sea. Gaping cracks across the flank's sloping surfaces stand as easily recognizable testimony to these violent tremors, most notably in a 2-mile-wide east-west zone called the Koaʻe fault system and in a long east-west series of giant seaward-facing cliffs, called *pali* in Hawaiian. Some of the cliffs are 1,500 feet tall. These tall headwalls of huge landslides look like irregular risers in a giant, chaotic staircase that descends to the sea. The builder apparently was untrained or inebriated.

Three historic earthquake-caused landslides—in 1823, 1868, and 1975—have added tens of feet to the height of some of the ill-designed risers. Presumably, a myriad of earlier temblors and associated ground slippage accounts for the remaining height of the lofty seaward-facing pali. As long as the volcano remains active, existing pali will continue to grow and new ones will form.

Volcanic hazards on the south flank are very real, even if they are not of the hot magma and lava flow type. The large south-flank earthquake in 1975 and its associated landslides generated a tsunami that swept across a campsite along the south coast of Kīlauea, killing campers. The desire to chart and forecast the process of earthquake-generated landslides certainly is one reason, and a powerful one, that HVO exists.

A growing and shaking volcano creature, then, was to be my laboratory for the next three years. The fundamental question asked of me and the rest of the science staff was just how the head, arms, and stepped torso became what they are today. The million-dollar question for us was to correctly forecast what the volcano will develop into tomorrow.

An eruption was in progress when I joined the staff at HVO, in August 1969. And true to the typical experience, I was to witness much more eruptive activity in the weeks and months to come. I experienced none of the frustration that would have come from living three years on a volcano that slept soundly. Instead, nearly nonstop eruption, at three different sites on Kīlauea, would be on my professional menu during the following three

Steam rises from hot lava as a lava fountain from Mauna Ulu feeds streams of molten rock cascading into ʻĀloʻi Crater. —Wendell Duffield photo, U.S. Geological Survey

years. The lessons we learned from these eruptions resulted in a quantum step in understanding the volcano's behavior.

I quickly became addicted to Kīlauea and frustrated by failures in forecasting what the volcano would do next. Toward the end of my three years at HVO, I was so addicted that Kīlauea occupied my thoughts almost as much as Anne and Mingo. While my live-in family members were always most important in my life, the specter of unpredictability in Kīlauea haunted my thoughts.

Meanwhile, at 120 quarantine days, Mingo was out of his lockup and happily ensconced in our Volcano house. With immediate eruptive action and lots of geology to discover, I was ready to hit the ground running.

Volcano Doctor Hans Schmincke approaches his feverish Mauna Ulu patient cautiously, 1969.
—U.S. Geological Survey photo

7

TAKING THE PULSE OF KĪLAUEA

An active volcano resembles a living animal in several instructive ways. The formal geologic categories for the stages of a volcano's life—active, dormant, and extinct—even carry connotations of the life stages of an animal. From the small beginnings of the very first eruption, an active volcano incrementally increases in size through time. With changes in size come changes in shape, some slow and subtle and others rapid and dramatic. Volcanoes breathe at places called fumaroles by exhaling magma-released gases, and the flow of a volcano's breath is all outward. Eventually, even the largest of volcanoes stops breathing and growing, and then through erosion follows the animal fate of ashes to ashes, dust to dust.

Volcanologists sample, measure, record, and then try to make sense of volcano-life processes, much in the way a medical doctor addresses a human's life processes. At Kīlauea in the late 1960s and early 1970s, our principal instrumental methods of examining the patient were precision leveling, electronic distance measurements, and seismometry. Studies of volcanic exhalations were in their infancy. Direct visual observations and laboratory studies of Kīlauea's lava rounded out our chest of tools.

Today, as geologists sip coffee in the office, a satellite accomplishes some of the leveling and many of the distance measurements that geologists used to do themselves. By comparison, my experiences might be thought of as "the good old days," or at least the "old days," when such surveys required the well-coordinated fieldwork of a team.

The technique of leveling we used resembled the one a woodworker employs when using a carpenter's level. Our surveying telescope carried a similar bubble level that allowed us to level the device precisely. By sighting through the carefully leveled telescope, we could determine the difference in elevation between two points. Circular brass plates, called benchmarks, cemented into a solid rock at known elevations served as reference points. Surveyors commonly report elevation relative to mean sea level, so we linked our survey to a tide gauge at Hilo Bay.

Working on an active volcano, we were primarily interested in changes in elevation because they indicate the internal restlessness of the live volcanic patient. So we determined the elevation of a point, or set of points, at different times to see if the ground had moved up, down, or remained stationary in the interim. Depending on weather conditions and the distance of the line of sighting from telescope to benchmark, we could measure a change in elevation of less than one-tenth of an inch. Experience has demonstrated that the surface of an active volcano can rise or subside many inches or even several feet in a matter of months, days, or even hours. Leveling, then, is a powerful tool for taking the pulse of Kīlauea and its ilk.

Electronic distance measuring, or EDM, is a sort of twentieth-century version of a yardstick—a tool to measure length or distance. With the advent of lasers, EDM instruments have been built to very precisely measure distances up to several miles. The instrument is set up over one benchmark, and the instrument's source of laser light is aimed at a reflector set up over another benchmark. The distance between the two benchmarks is a direct function of the travel time of light from laser to reflector and back to laser, with reference to the wavelength of the particular light. The EDM instrument, called a Geodimeter, in use at HVO while I was there could measure as precisely as one part per million, which translates into knowing the distance between two points that are about one mile apart to within a small fraction of an inch. The Geodimeter, then, could detect minute changes in distance between two benchmarks, making it, like the telescopic level, a powerful tool for studying an active volcano. The combined results of EDM and leveling surveys define both lateral and vertical movements of the ground.

A seismometer puts the physicist's concept of inertia to work in a very practical way. The key idea is simply that an object at rest (or in any given state of motion) tends to stay at rest (or in that state of motion) unless acted upon by some outside force. The seismometer must be the ultimate lazy man's tool, for its job is to stay at rest, essentially doing no work of its own. However, should the earth beneath a seismometer happen to move, as during an earthquake, an immobile pen on that instrument makes a tracing of the earth's motion by staying at rest while a chart firmly and directly attached to the earth moves back and forth beneath the pen. Pens come in a variety of ink-filled and electronic forms. In the end, when the earth is once again still, the scientist has a graph of the earth's motion during the quake, recorded within a very precisely timed framework. Earthquakes tell us when and where magma is moving in and through the volcano, and when parts of the south flank slip toward the sea.

So, with our arsenal of telescopic levels, Geodimeter, seismometers, hammers, sample bags, gas sniffers, and eyes, we set out to learn what we could about Kīlauea's behavior. Understanding never seemed to come easy, but we learned that a combination of teamwork, persistence, and patience could pay handsome dividends.

Teamwork was one of the most important elements of our efforts, for no individual could accomplish the tasks at hand. When seen together, though, we hardly resembled a team. We numbered about fourteen, as temporary helpers would come and go. The permanent staff—Akira, Arnold, Bill, Bob, Dallas, Don, Don, George, Jeff, John, Ken, Maurice, Marie, Reggie, and me—included representatives of both genders, multiple ethnic groups, strongly diverse educational backgrounds, and ages from nearly geriatric to wet behind the ears. All but four team members—Don, Don, Dallas, and me—were from the local Big Island population. Toward the end of my three-year stint at HVO, Chris and John, two more volcanologist trainees from the mainland, replaced Don Swanson and Dallas. The characters mentioned in this collection of essays, though, come entirely from the team before Don and Dallas left.

Our "coach" was senior scientist Don Peterson, who had the daunting task of trying to satisfy two sets of bureaucrats, those above him in the U.S. Geological Survey and those in whose National Park his observatory stood, while simultaneously helping facilitate the work of the entire HVO staff. Moreover, while his coaching role was at the top of the local U.S. Geological Survey organizational pyramid, we teamsters at the lower levels were sometimes only semicontrollable, though always in a friendly and constructive way. We worked out of ragtag quarters, a small building on

HVO staff during most of the author's tenure there (from left to right). Back row: *Maurice Sako, Glenn Nakamura, Ken Honma, Dallas Jackson, John Forbes, Don Peterson, Akira Yamamoto.* Front row: *Bob Koyanagi, Marie Onouye, Arnold Okamura, Wendell Duffield, George Kojima, Reggie Okamura, Don Swanson. Missing: Jeffrey Judd and Bill Francis.* —John Forbes photo, U.S. Geological Survey

the rim of Kīlauea Caldera. It probably was, and is, no accident that HVO is antipodal to the Volcano House Hotel and its elegant guests, across the 2-mile-wide caldera.

By chance and good fortune, our team was about as spiritually and mentally compatible a group imaginable, which made much of the field work enjoyable as well as scientifically rewarding. We found ways to make repetitious and boring tasks into friendly and spirited competitions, but competitions where everybody won. We often seemed to be having more fun during our fieldwork than many of the National Park tourists we met in the path of our surveys. These memories of camaraderie are a large part of what motivates me to write about my HVO days, nearly thirty years later.

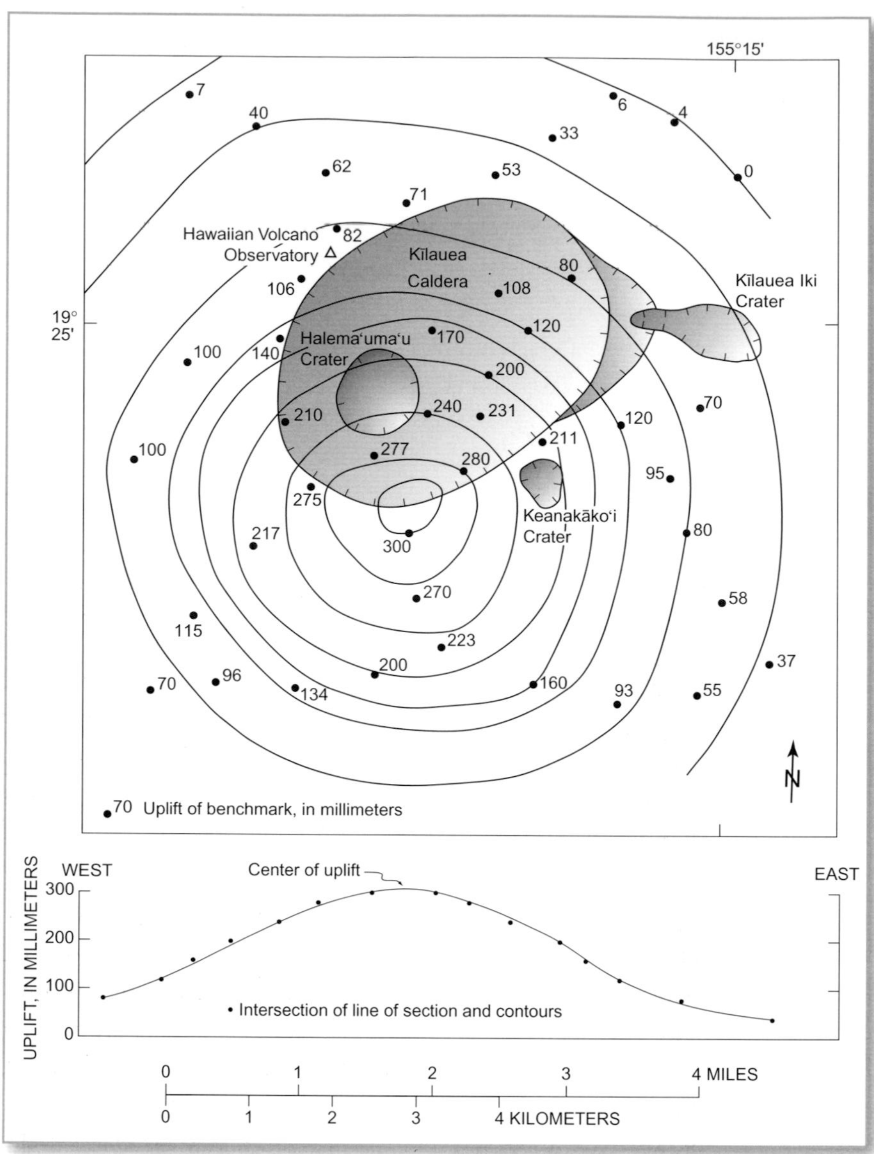

Map and cross section of ground doming at the summit of Kīlauea during a four-month period in late 1971 and early 1972. The numbers are measurements, in millimeters, of the surface's elevation gains—the greater the number, the greater the bulge. (There are 25.4 millimeters in 1 inch.) The bull's-eye pattern of uplift shows that the ground rose nearly 12 inches (300 millimeters) during the four-month period of inflation. During that time, Kīlauea was gorging itself with magma, which it spewed out in an eruption that began in February 1972. I can't honestly say we accurately forecast this eruption from the results of the leveling survey, but the growing domelike inflation blister held our close attention as eruption day approached. —From Duffield and others, 1982a

8

ARE YOU ON THE LEVEL?

Over periods of days to months to years, the summit area of Kīlauea Volcano behaves much like the top of a large spherical balloon. When magma enters the volcano from its deep mantle source below, the volcanic mountain expands, or inflates, to make room for this new material. When magma exits the volcano during eruption, the mountain contracts, or deflates. If enough magma erupts rapidly, the solid roof of the mountain drops in pluglike fashion to enlarge the caldera basin. It's no surprise then that our leveling surveys typically indicated the center of the balloon sat beneath the caldera. Though an oversimplification, the balloon analogy is a useful way to organize one's thoughts about volcano behavior. The balloon concept is the first physical model HVO scientists developed from their surveys of ground deformation, and it has withstood the test of time remarkably well.

One task of the scientific staff is to periodically determine the degree of inflation of the Kīlauean balloon and to somehow translate its measured

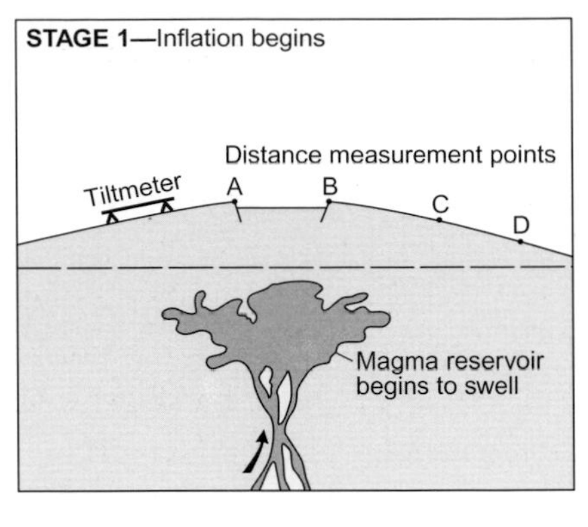

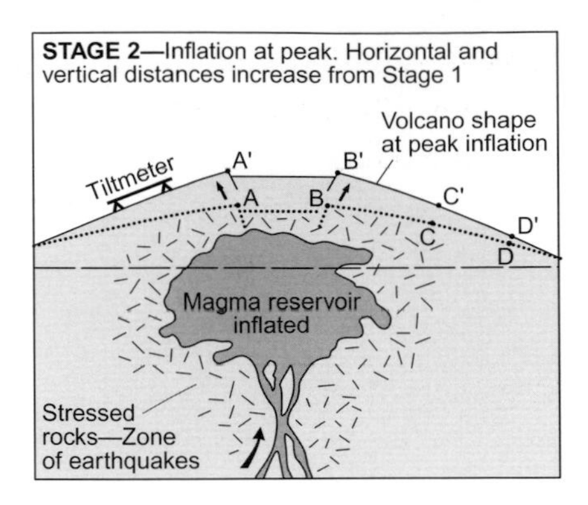

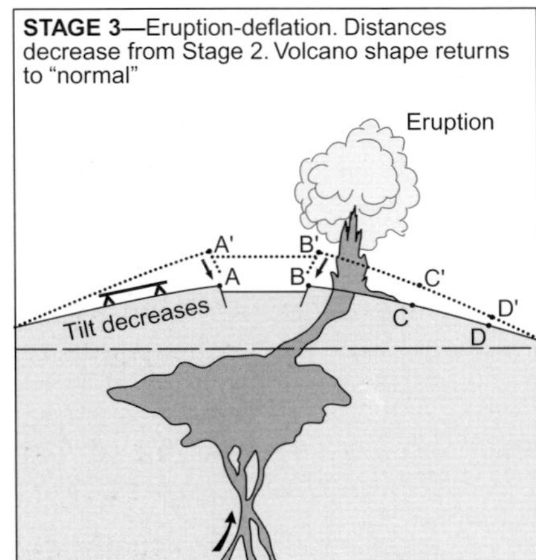

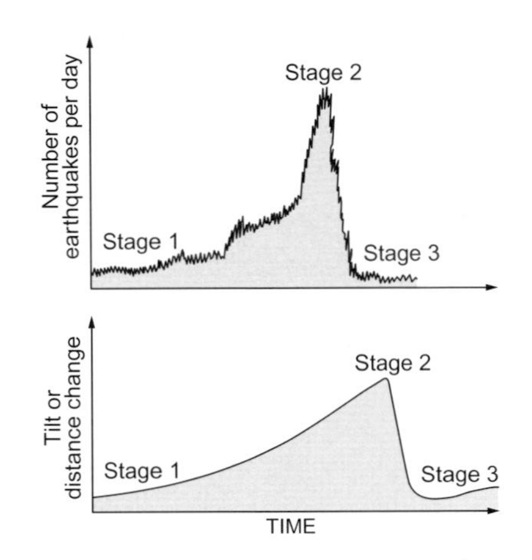

The three diagrams illustrate stages in the inflation and deflation of Kīlauea as magma from the mantle enters and then erupts from the volcano. The two graphs show how earthquakes, ground tilt, and EDM distances change through an eruption cycle.

condition to a forecast of when the magma-gorged balloon might burst into eruption. The original, and still the principal, technique for determining the state of inflation is leveling.

Unless specific volcanic events indicated the need for more frequent measurements, we did our leveling at about three-month intervals. This was often enough to keep tabs on Kīlauea's inter-eruption swelling, but not so frequent to cause staff burnout. To level the summit area took the efforts of most of the staff for about three consecutive days. Ideally, such a survey should be completed instantaneously. But we were only humans with limited instrumental resources. So, to minimize any inflation or deflation that might occur during the survey, we worked continuously over the three days until the job was done. Someone stayed at the office to answer the phone, but otherwise HVO was nearly vacant and very quiet during the days of leveling.

A map of the survey network is a hodgepodge of interconnected lines and loops that crisscross a roughly 5-by-5 mile patch of ground centered on the caldera. Though this network might initially look like the random wanderings of a child's crayon, we established the lines to cross key geologic structures and to take advantage of as many existing roads as possible in the process. The network was cumulatively many miles in surveyed length.

One line started—or ended if you looked downhill instead of up—at Hongo Store in Volcano, a location far enough from the caldera that we could reasonably consider its elevation unchanged between leveling surveys. This was our nominal zero when it came time to contour elevation changes. However, if disruptive volcanic events or the passage of too much time argued against such stability, we leveled an additional 26 miles to a tide gauge at Hilo Bay, an effort that even our happy-go-lucky team was never eager to repeat. The extra miles and the elevation increase from sea level to about 4,000 feet doubled or tripled the time needed to complete a leveling survey.

Typically, a leveling crew consisted of four people. Sometimes a fifth person came along to replace anyone who needed a brief rest and to help maintain a robust level of our customary inane chatter. We usually ran two crews simultaneously in different parts of the overall network. Within a crew, one person "ran the gun," which meant setting up and operating the leveling telescope. Two others were "rodmen," each carrying a 12-foot-long stadia rod to be held vertically on the benchmarks. The rods, marked in feet, inches, and fractions of inches, are made of a metal that does not expand or shrink as the temperature changes. This material guarantees that a foot on the rod in warm sunny weather is still a foot on the rod in cool and cloudy weather. The gunman aimed and leveled his telescope on a distant rod, then peered through the eyepiece to read where the telescopic crosshair fell on the graduated rod. The fourth person recorded numbers that the gunman called out.

The distance between gunman and rodmen varied with the slope of the land being surveyed but normally was around 100 feet or a bit less. Once the gunman had completed a backsight on rodman number one, he spun the telescope about 180 degrees on its vertical axis, refocused, and began the foresight on rodman number two. Meanwhile, rodman one

Wendell Duffield sights through a leveling telescope to a stadia rod and shouts readings to a note taker, who also drives the car from station to station. —HVO staff photo, U.S. Geological Survey

moved ahead to a new benchmark and, once the gunman moved forward, became the next foresight. In this manner, the crew leapfrogged along. On roadways, the person who recorded data also drove a car. The gunman usually hitched a ride from setup point to setup point. When leveling cross-country, everyone had to walk. On or off road, rodmen typically jogged from benchmark to benchmark, especially when interteam competition was under way.

Obviously, the field process of leveling is repetitive and potentially very boring. We kept the job more fun than funk—as much reveling as research—by competing with each other. One competition was to reduce the time needed to complete the entire survey. Given the vagaries of weather and equipment, it was tough to ever finish in less than three days. We garnered more immediate satisfaction from minimizing the time it took to survey a mile of line. This competition became particularly fierce along roadways, where the gunman rode from one setup point to the next while rodmen ran from benchmark to benchmark. The winner enjoyed bragging rights only, but those were enough to motivate us.

Though speed added spice to this otherwise bland work, we never sacrificed precision or accuracy for swiftness. Professional surveyors should be impressed that our results qualified always as "second order" and sometimes as "first order" leveling. It doesn't get any better than first order. And thank goodness for such an active volcano—the ups and downs of Kīlauea are commonly so substantial that even sloppy leveling would have captured the main pattern of elevation change.

Beyond our competitions, we had fun in our interactions with the public. Many a curious and gregarious tourist would stop to chat. Most approached us with the preconceived notion that we were part of the advance crew for a road-building project. We impressed everyone when we replied that, no, we had nothing to do with road construction. On the contrary, we might be involved with road destruction, because we were measuring the volcano's magma-induced swelling, which was going on right underfoot. This news usually resulted in amused disbelief or sudden concern for safety and a quick retreat to a departing vehicle.

Then there was the time we masqueraded as a prison gang. On the suggestions of some National Park rangers, who were required to wear uncomfortable-looking uniforms day in day out, we decided to take on a formal team identity. The Park rangers and HVO staff were all employees of the Department of the Interior, albeit of different bureaus in that department. And an unspoken yet palpable attitude coming from some Park Service personnel communicated that we U.S. Geological Survey folk were a visual embarrassment to the Department of the Interior in Hawai'i. These Park people may have been right.

In a good-natured attempt to simultaneously gently needle and placate our fellow Interiorites, each of us bought a pale blue cotton shirt of a quality that could never be mistaken for Brooks Brothers. On the back of each shirt, we stenciled "O'ahu State Prison" in bold black print. Below this name we added a typical several-digit, inmate-identifying number. These shirts, combined with our usual shabby attire and unkempt visage, prompted more than one tourist to comment about the value of getting useful work from prisoners rather than letting them loaf behind bars. And we were indeed prisoners, totally captivated by the wily Kīlauea Volcano.

9 A LOAF OF BREAD, A JUG OF WINE, A GEODIMETER, AND THOU

If leveling smacked of drudgery—interteam competitions notwithstanding—geodimetering was akin to painful death. As though hours of intellectually stultifying repetitiveness weren't punishment enough, a body applied to geodimetering usually suffered physically in the process, and in isolation from commiseration. However, I found one particular assignment so delightful that I requested the job of reflector man whenever a Geodimeter survey rose to the top of the "to do" pile.

Ideally, level and Geodimeter surveys were completed in rapid succession. The need for speed was twofold. First, if we knew both vertical and horizontal changes "instantaneously," we then knew the complete pattern of volcano deformation, rather than just part of the picture. Second was the ever-present concern that some deformation might occur during the time it took to complete a survey. We always hoped Kīlauea would flex its body slowly and sluggishly while we quickly completed our surveys; our results tended to support that hope.

A Geodimeter crew consisted of the laser-gun man and his recorder, plus a complement of up to three reflector men. As in the leveling network, some survey stations were accessible to automobiles; others were not. To accommodate remote stations, we transported the Geodimeter in a backpack, but it weighed about as much as a normal adult could carry without fear of permanent dorsal deformity. The recorder was put to the acid test, literally, by carrying a 12-volt lead-acid battery, which provided the energy for the Geodimeter. Patterns of circular splash holes that corroded through the backpack and favorite surveying garb of the recorder paid naked testimony to the impossibility of climbing a cinder cone in the rain forest without sloshing some battery acid.

Once the laser-gun man and recorder had centered the Geodimeter over a benchmark, they measured the distance to as many reflector-occupied benchmarks as they could see. So while the initial set-up generally was strenuous for the Geodimeter crew, once set up, they stayed put for several measurements. A reflector man, on the other hand, had to scramble from site to site as quickly as possible for the efficiency of the survey. Fortunately, a set of reflectors was light and easy to carry in a backpack. Still, reflector

Having lugged the Geodimeter and associated equipment to the top of a cinder cone near Kīlauea Caldera, the laser-gun man shoots to several mirror stations while the recorder (Wendell Duffield) takes down the measurements. The south flank of Mau̦na Loa is faintly visible in the background. —HVO staff photo, U.S. Geological Survey

work typically was stressful, as the reflector man always raced against time marked by approaching cloudy weather. Without line-of-sight visibility, measurements were impossible.

Our sighting tools were mirror signals and two-way radio conversations. The mirrors established a line of sight, and the sun was the source of light needed for this technique to work. By knowing approximately where the other end of a line was, both Geodimeter and reflector men could scan a small sector of the landscape for that telltale bright flash from the other person's mirror.

Even in clear weather, once set up at their respective stations, the Geodimeter and reflector men sometimes had difficulty finding each other. The challenge was to be able to aim a tightly focused laser beam at a target less than 1 foot wide and 1 to as many as several miles away. For relatively short distances on clear days, this was a surmountable task. However, for long shots on partly cloudy, and even clear days, establishing the line of sight was difficult to impossible. Sometimes we simply had to give up, usually because clouds had moved into the line of sight, and return later during clear weather.

Radio chitchat alone was of limited use for helping the laser light find its target. The side of a volcano can look remarkably uniform from a distance of a mile or more. Rather, we used radio talk mainly to establish when a shot was about to get under way, was under way, and was successfully finished. Thus, radio communication was absolutely essential because it told a reflector man when he should be paying attention to his business, rather than waiting for his turn in the succession of shots from a particular setup for the Geodimeter.

In general, once the team successfully completed a shot, the reflector man hastened to his next station, in the never-ending race against cloudy weather. One exception, however, was a station known as Keakapulu on the Kapāpala Ranch. This place was so remote that it alone represented full duty. There was no need, or use, for the reflector man to rush off once the team finished the Keakapulu shot.

The drive to Keakapulu was long and rough, which discouraged most of my colleagues from lusting after the assignment. I didn't mind the bumps, though, and thoroughly enjoyed the perks, or at least what I viewed as perks that came with being Keakapulu man. I took this assignment whenever possible. If this was work, I wondered why anyone would pay for the privilege to play.

Keakapulu sits at 4,000 feet above sea level on the south-facing slope of Mauna Loa, about a dozen air miles northwest of Kīlauea Caldera. This station carried special importance to the overall Geodimeter network because Mauna Loa was in a period of dormancy during my HVO years and, thus, served as a stable reference frame, in the same way that the benchmark at Hongo Store in Volcano was a stable reference for our leveling surveys.

Morning clouds tend to bank against Mauna Loa far more often than not, so a Geodimeter shot to Keakapulu was an early-morning effort, the first attempted for the day. Sometimes, with patience, we could outwait the weather and catch a break in clouds long enough to complete a shot later in the morning. More commonly, though, I drove to Keakapulu, waited in vain, and returned to HVO in abject failure. Abject at least for the professional side of such a day. Elation was always the personal norm for a Keakapulu trip.

Anne, our dog Cinda, and I made Keakapulu days into family picnics. To get to the station shortly after sunrise, we had to leave HVO about the time sun first thought about rising. Such necessary early departure put off most of my colleagues from Keakapulu duty. Once under way, though, the drive offered a kaleidoscope of Big Island pleasures.

The first several miles traversed pavement, downhill to the southwest on Hawaiʻi 11, toward Pāhala and the Kona Coast. From near a lovely National Park campground, Nāmakani Paio, nestled in a grove of huge eucalyptus trees, the drive begins across smooth to ropy surfaces of lava flows. The Hawaiian name for this type of lava, pāhoehoe (say PAH-hoi-hoi), is pure onomatopoetry. A mile or two farther, the road moves onto the spiny and rough surfaces of lavas called ʻaʻā (say ah-AH)—another example of the Hawaiian language's facility to coin a word that sounds like its object. The road stays on this rough stuff until the turnoff to Keakapulu. Though virtually no soil covers the ʻaʻā, a forest of ʻohiʻa trees decorates the landscape, growing from rootholds in impossibly infertile-looking bare rock. Abundant rainfall apparently substitutes well for soil. In season, these trees carry bottlebrush-like red flowers, called lehua, with the occasional yellow variety for interest.

Then comes the turnoff from Hawaiʻi 11 onto a rough track across pasture and more lava fields. The rough ride requires us to reduce our speed, which I see as enhancing our opportunities to look at the landscape. This glass is definitely half full.

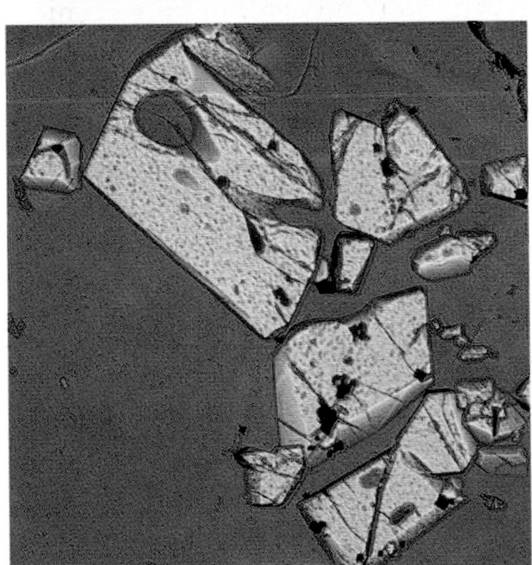

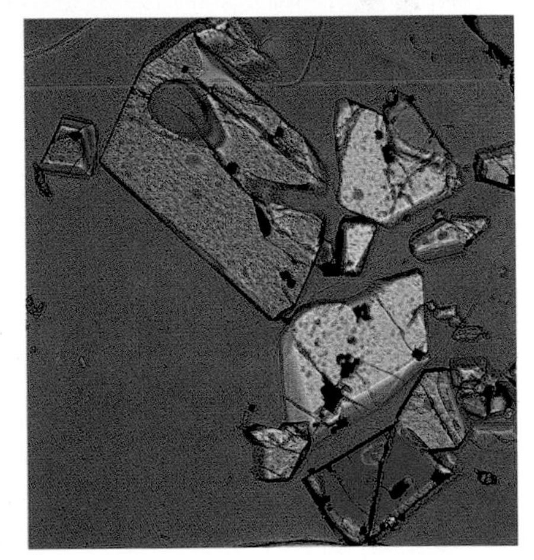

Top: *Geologists call the smooth-surfaced variety of lava, as seen in this yard-wide front of a lava flow at Kīlauea, pāhoehoe. Lava with a broken and irregular surface* (not shown here) *is called ʻaʻā.* —Wendell Duffield photo, U.S. Geological Survey
Bottom: *Both varieties of lava look similar in a very thin slice of rock seen magnified four times real size. The uniformly brown part of the rock is glass, and the multisided grains are crystals of the mineral called olivine, which grew in the magma before eruption. Ordinary light illuminates the sample at lower left; polarized light illuminates the sample at lower right and produces various colors in olivine grains.* —Rosalind Helz photos, U.S. Geological Survey

At this point, we are outside the National Park and climbing the south-facing flank of Mauna Loa, across land used principally for grazing cattle. A carpet with wind-blown grass covers areas free of trees. Flowers and the occasional guava bush interrupt the grassy flats. The truck scares up a variety of birds, whose names I never did learn, as we bounce ever upward toward Keakapulu. Nēnē, the Hawaiian goose and about the only Hawaiian bird name I can remember, once lived on these slopes and was reintroduced just a few years after my time in Hawai'i.

The pitch of the road alternately increases and decreases, but it always heads upward. We are ascending Mauna Loa's giant staircase, a series of pali or headwalls of huge landslides. These pali are older and so a bit less abrupt than those on the south flank of Kīlauea, but otherwise record a similar history.

On one of the gentler rises, we pass by a tumbledown cabin and the remnants of a rock wall, built in a square about 500 feet on a side. Was the wall meant to keep cattle in or out, we wonder. Whoever once lived here had planted eucalyptus and koa trees, now hundred-foot-tall sentinels at irregular spacing along the wall.

Finally comes the last steep pitch, which brings us to a broad grassy bench called Keakapulu Flat, the site of our reflector station. I center the

Anne and mirrors at Keakapulu Flat—Wendell Duffield photo, U.S. Geological Survey

reflector on a tripod over the benchmark and point it toward the summit of Kīlauea. I radio Don, the Geodimeter guru for the day, to let him know that I am in place and ready. And I wait.

Meanwhile, Cinda spirals around in doggy delight, and Anne has laid out blanket, snacks, and beverage. It's never too early in the day for a picnic at Keakapulu. Anne starts reading a book.

We are totally alone. Not even cows seem to inhabit this supposed pasture. At this point, the rest of the morning will follow one of three courses.

Course one is in cloud-free sky and has the Geodimeter shot successfully completed within a few minutes of our arrival. Being Midwesterners easily riddled with the guilt of a Protestant work ethic, we have an abbreviated picnic, pack up, and head back down the mountain. We enjoy the drive, both up and down, but forego the lingering picnic treats.

Courses two and three include waiting and hoping for the usual morning buildup of clouds to disperse long enough to complete the shot. Sometimes the wait pays off and sometimes it doesn't. We need only about fifteen or twenty minutes of clear visibility. Anne, Cinda, and I are always in sun, because the unwanted clouds hover somewhat below us, in the direction of Kīlauea Caldera.

While I wait, Don is completing shots to other reflector men, who are scrambling from station to station down around the summit of Kīlauea. Simultaneously, both Don and I should be watching the Keakapulu weather. We will interrupt one of the easy-to-get shots near Kīlauea Caldera in midstream if Keakapulu clouds disperse.

The difficult part for me is paying attention to cloud cover, when so many sensory temptations lie within reach. Quite literally, the loaf of bread, the jug of wine, and Anne are right there on an inviting blanket. Cinda entertains herself.

With such distractions, did I ever fail to notice cloud dispersal or perhaps fail to respond promptly to a radio call from Don? Don or his recorder Reggie should answer these questions. Experience taught Don and me that we likely waste time by trying to outwait the weather beyond a couple of hours. Experience taught Anne and me that a couple of hours were adequate for picnic delights. Shot completed or not, I always left Keakapulu reflector duty with a smile on my face.

Snake-finder Cinda peeks between ferns of the rain forest in Volcano. —Wendell Duffield photo, U.S. Geological Survey

10

YOU'RE RIGHT! THAT IS A SNAKE!!

The entertainment industry generally portrays tropical rain forests as dreadfully dangerous places to visit. If you are not first sucked into a mire of quicksand, you almost surely will be savaged by some conspiring group of human-devouring plants and animals. You certainly are not safe sitting idly, even briefly, where the relatively immobile killers can wreak their havoc on human flesh. And motion is no guarantee of safety because so many animals can outrun or outlast a fleeing person. Hollywood portrays snakes as particularly insidious—they blend in with the tropical vegetation and can move swiftly and quietly to overtake their human prey. If even a dollop of this popular scenario were true, why would anyone want to traipse cavalierly through the Hawaiian rain forest in search of the holy volcano grail?

Luckily for local residents, tourists, and volcanologists alike, Hawai'i is devoid of snakes. At least, that is what the Hawai'i Visitors Bureau and

Chamber of Commerce advertise. And that is what all locals "know" to be true. Imagine, then, the statewide surprise, shock, and clamor when Anne and I discovered the Islands' first snake, loose and free to slither wild in the rain forest of Volcano. As a discovery of almost religious proportions, it seems appropriate that we found the snake on a Christian day of worship.

Every Sunday morning, weather permitting, we walked our dog Cinda on a new and circuitous route along the roads of Volcano with the eventual destination of the Okamura Store, where we bought a copy of the local newspaper. Following the usual chat with the proprietor and neighbors at the store, we walked home to sip Kona coffee as we read the comic section and current news about the Big Island. Because of Cinda, one Sunday walk made statewide news of its own.

Not wishing to disturb U.S. Geological Survey bureaucrats by suggesting more than a cat would be living with us in Hawai'i, we had acquired Cinda after moving to Hawai'i. She was another Anne-the-veterinarian project and a well-trained, full-fledged member of the family. She also fulfilled a recommendation from the police who had investigated the two burglaries of our house on the mainland shortly before we moved to Hawai'i.

The first of those burglaries bred discouragement enough, being the first such violation that either Anne or I had experienced. The second break-in and robbery, though, just a few weeks later, drove us to plead the police for ways we could enhance home security beyond the usual locks, bells, and sirens. Their advice was quick, short, simple, and not reassuring for anyone seeking safe living in a solid suburban neighborhood: move or get a large guard dog. Though not in direct response to this advice, we did both.

Cinda was a Doberman pinscher. We would have purchased a Doberman in Hawai'i, even without a police recommendation. We were especially fond of the breed's sleek lines and intelligence (read trainability here). Cinda excelled at her obedience-training class in Hilo, and we could take her for walks without worrying that she would set off on her own itinerary and with her own agenda.

So, as we strolled down a Volcano road that Sunday morning, we knew something was amiss when she refused to come when given that command. Our blood pressure up a bit from this unusual show of disobedience, we both walked back to where Cinda had stopped along the shoulder of the road, her attention riveted on a particular spot in the long grass. She was barking furiously. When we saw what had her so transfixed, our blood pressure went even higher. Coiled loosely there in the grass lay an unmistakably large and beautifully patterned snake. My instant reaction told me that the creature was about 12 feet long, with a diameter at least that of my upper arm or perhaps my thigh. This was not a normal Sunday stroll.

Contrary to the Hollywood image of rain forest dangers, this snake in the grass did not attack us. Though we could see no signs of injury, the elegant creature seemed barely able to move, even in response to our gentle prodding. Apparently, the chill of a 4,000-foot-high Volcano morning was enough to immobilize a cold-blooded serpent.

Anne and I both appreciated the implications of our find here in this purportedly snake-free paradise called Hawai'i. Unless the Chamber of Commerce story about the absence of serpents in Hawai'i was a hoax, we had something unique on our hands. The question was what to do without losing the living evidence of our discovery. A quickly hatched plan left me along the road to keep tabs on the lethargic snake, while Anne, with Cinda, headed back to our house to phone Win Banko, a friend and research biologist living in the National Park.

I learned later that Win was sure Anne was playing a practical joke, or maybe under the influence of some mind-altering drug, when she called. Like everyone else in Hawai'i, Win knew there were no snakes on the Islands. And so he argued long and with that force of logic that comes from knowing you are right. He even tried to convince Anne that we had happened upon a mongoose realistically imitating a snake. After a protracted phone conversation, though mostly out of friendship I suppose, he agreed to interrupt his Sunday-morning ritual and come see what Cinda had found.

When Win arrived at my roadside post, chilled serpent lying still in the grass, just a quick glance brought forth a startled "You're right! That IS a snake!!"—which, of course, Anne and I had known all along. After all, it doesn't take a trained biologist to recognize such an easily identifiable creature, just as it doesn't take a trained geologist to recognize a rock.

Win, though, jumped quickly beyond the snake label to thinking this was a python. (The beast was later identified as a reticulated python.) A python, Win noted, when full-grown would be about 30 feet long and appropriately larger in diameter. This nugget of biological wisdom suitably impressed us.

Although he hadn't really believed Anne's story, Win had come prepared. He had a wire-mesh cage and had quickly improvised an adjustable loop on the end of a long pole, the kind of tool professionals use to pick up snakes when hand contact is ill advised. He teased the snake into raising its head, which he snared in the loop. As he deposited the regal snake in the cage, he held it vertically long enough to estimate a length of 4 feet. Still chilled, the snake rested as quietly in the cage as it had in the grass. Win drove off, and we never again saw Cinda's discovery.

We certainly heard about the snake, though.

Win promptly and responsibly notified state officials, whose immediate reaction was to destroy the animal. Though the python was securely caged, the call for its destruction seemed to stem from a fear that worse evil than that perpetrated by the serpent in the Garden of Eden would follow from the mere presence of a "wild" snake in Hawai'i. After all, such a feeling so deeply permeated official state thinking that law prohibited keeping any snake, except at the Honolulu Zoo. Even there, only single snakes were allowed, to avoid the possibility of serpent fornication and the expectable consequences thereof. Who could blame Hawai'i for wanting to stay snake free?

No pun intended, but destroying the snake at this point seemed overkill to us. If baby snakes had hatched somewhere during python's presence in Hawai'i, it was too late to capture them by killing Mom. Win convinced state officials to drop the death plan, on the condition that the snake would be shipped off the Big Island, and away from the state, as soon as possible. Colleagues on the mainland helped make arrangements to have the snake received at the National Zoo in Washington, D.C.

Being a creative researcher and educator, Win used the interim productively. On the drive to the airport, he took the snake on a tour of schools, much to the unbridled delight of students. Many of the students had never before seen a live snake because state law excluded snakes from importation and many of these children had never traveled beyond the bounds of their state, or even their island. To see such a large, beautiful, and unusual animal was a treat as tasty as four-day poi. And being well fed by Win, the python was docile and friendly, even at the blood-circulating temperatures typical of schools near Hawaiian sea level. Before winging its way to research associates at the zoo on the mainland, the python visited many Hawaiian public schools in only a few hours.

Reports of the entire saga of the snake hit the media—radio, TV, and newsprint alike. Anne and I stayed in the background of the various news stories. Cinda got some public exposure and recognition. But the stories focused mainly on the snake and what its presence might mean for life on the Big Island. No one likes to lose something of value, and Hawaiians feared that part of their paradise had been compromised. The emotional question "how could this happen?" was in search of an answer, and if an important piece of paradise truly was lost, no answer would seem adequate.

Adequate or not, Anne and I speculate that Hawai'i was a victim of war, the Vietnam War in this case. Fighting in Vietnam was under way when Cinda made her discovery. Moreover, Hawai'i Volcanoes National Park is home to Kīlauea Military Camp, and KMC was a logical and popular destination for soldiers on leave from the fighting. For decades, this place of rest and recreation has been popular with military folks, during war and peace.

In our version of the snake's arrival, a soldier came to visit KMC with a Vietnamese snake illegally in tow, perhaps literally up his sleeve. When he became bored with the animal, he released it into what seemed the snake's natural habitat. If such a soldier had not been Hawaiian, which is very likely, he might even have reasonably viewed the release as a favor to the snake and certainly as causing no harm to the rain forest. A mainlander imbued with effective Hollywood training knows that a rain forest, any rain forest, comes with many built-in friends and potential mates for a snake.

Win recalls a story, originating with local state officials, about a Big Island resident who may have somehow had the snake illegally imported.

However, the suspect denied the alleged crime, and no charges were brought against him. The mystery remains.

However a young python came to be in Volcano that fateful morning, I never again chased lava through the rain forests of Kīlauea with the same carefree feelings of safety from animal attack that I had before Cinda's discovery. As far as I know, though, no other snake has been found loose on Hawai'i since. As my Hawaiian friends liked to say about almost any situation during my HVO years, "Hey brudda. It's no big deal." With all the tromping around we do in the rainforest during our field investigations, I sure hope they're right.

11

YELLOW RUNOFF

Futurists and pessimists sometimes say that insects will inherit the earth. I interpret such dire prognostication as another way of saying humans foul their nest so effectively that surely they will not be around much longer. Insects, on the other hand, seem to adapt well to environmental change. They have been around in one form or another for some hundreds of millions of years and will fill whatever niches we ephemeral humans leave.

The longer I live, the more I tend to believe that one, if not the, limiting factor to human life on earth will be an inadequate supply of potable water. I suppose time and events might prove me wrong—water covers three-fourths of the earth's surface, after all. But at a minimum, my crystal ball shows the price of potable water increasing at a rate greater than that for most other consumable goods. The inventor who perfects a way to desalt large volumes of seawater at a competitive cost is bound for great wealth.

Humans in Hawai'i have had to scratch for an adequate supply of potable water almost since their first arrival on the Islands. This situation seems contradictory for a place whose windward side receives hundreds of inches of rainfall annually. However, rainfall doesn't stay long on the ground in Hawai'i. People have to capture the rain as it falls or overtake the water in the great Hawaiian water chase as it moves underground on its way to the sea.

The Hawaiian Islands, including the Big Island, are built up almost exclusively of lava flows. Geologists call this lava rock basalt, which is the most common volcanic rock on earth. In a typical roadcut in the Hawaiian Islands, you will see the bulldozer-truncated edges of multiple lava flows stacked one atop the other, each one just a yard or two thick.

Even the most casual of observers will notice that a stack of lava flows contains considerable interconnected open space. Some of those open spaces are bubbles, called vesicles, that became trapped in lava as it cooled from a frothy and gassy melt to a solid rock. Other openings are spaces preserved between tumbled pieces of lava crust that broke up when the still-molten core of a flow continued to move after the crust formed. Still other open spaces are cracks that formed during earthquakes or when the lava shrank as it cooled and solidified.

With so much open space present beneath the surface, rainfall in Hawai'i almost immediately soaks into the ground and begins a downward journey deep into the pile of lava flows. The water travels until it reaches a depth at which earlier rainwater already saturates the basalt. The top of the zone of saturation is called the water table. Under large expanses of con-

An idealized, yet widely applicable, model for the occurrence of freshwater within a typical Hawaiian island

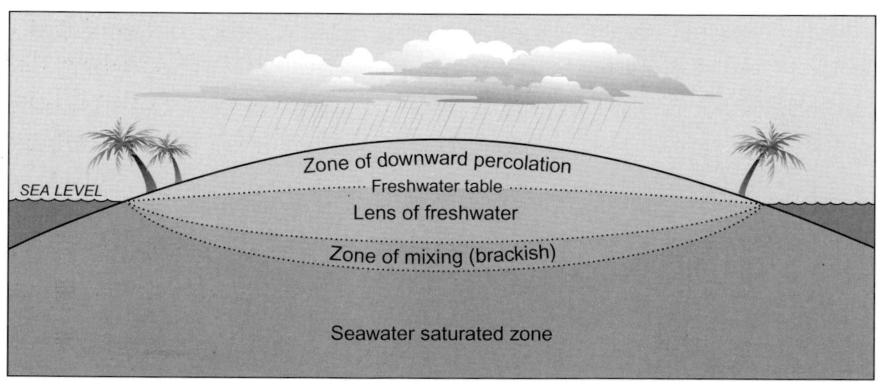

tinental areas, the water table is only tens to a few hundred feet below the surface, which favors the cost-effective drilling of wells for potable water. On the Big Island and similar volcanic islands, the water table is almost everywhere just a little bit above sea level, regardless of the overlying land elevation.

Seawater also seeps right through the pile of cracked and broken volcanic rock that makes up the island. With repeated rainfall, downward percolating water forms and replenishes a lens of fresh water that floats on the salty, and therefore denser, seawater at about sea level. The freshwater and the underlying salt water slowly mix, but as long as the current rate of rainfall persists in Hawaiʻi, so will a lens of water fresh enough for human consumption. In the long-term, though, we might want to consider that increasing human demand for potable water might someday outstrip the ability of rainfall to maintain the freshwater lens.

Given the configuration of the water table, almost all wells in Hawaiʻi are drilled into the freshwater lens from platforms near sea level. High on a volcano, drilling a well for freshwater is either prohibitively expensive or physically impossible. For instance, to reach the water table beneath the top of Mauna Loa would require a hole nearly 14,000 feet deep! Instead, Hawaiians use a cost-effective alternative means for providing potable water at the high elevations—they capture the rainfall before it goes underground.

Now you know why my rental house in Volcano, at 4,000-feet elevation, had a water-storage tank to catch rain that ran off the roof. All houses and tourist facilities in Volcano and in the nearby Hawaiʻi Volcanoes National Park had and still have that type of water system. The National Park Service at Kīlauea maintains a system that supplies potable water to staff and Park visitors alike. To do so requires an extensive and well-integrated system of plumbing that collects rainfall in storage reservoirs and then distributes it to consumption points. Though never discussed publicly, for reasons that will become evident, this system experienced a bizarre and colorful episode of plumbing-gone-awry during my HVO tour of duty. The episode began when repair was needed in the men's restroom at the Volcano House, the lovely tourist hotel perched on the rim of Kīlauea Caldera.

Ordinarily, a leaky pipe that carries wastewater from a restroom is easy to fix. Such a pipe typically is under little, if any pressure, because gravity gently yet quickly pulls liquid waste through a pipe to its disposal destination. In contrast, high pressure continuously and vigorously pushes on a water-delivery pipe. Thus, isolating part of a drain system for repair work is safe and simple—even an inexperienced squirt should be able to do the job without fear of an unwanted flood. If no liquid is added at the top, nothing will flow out the bottom. For the case in point, the workers isolated the offending pipe for repair simply by locking the restroom door.

However, the drain system at the Volcano House and other plumbed facilities along Kīlauea Caldera in the National Park is relatively complicated because it incorporates two gravity-driven piping systems: one for liquid waste and another for collection of rainfall from a building's roof. The former liquid is truly waste and thus disposed of in an appropriate manner. But the latter is pumped to large nearby tanks and then distributed back to the various buildings as potable water.

Confusion between the two gravity-driven systems of piping is possible, unless each is clearly marked along its entire and undoubtedly circuitous route through a building. Apparently, the pipes lacked such marking when I was working at HVO. When confusion reared its ugly head, one particular plumbing repair at the Volcano House routed male urinal outflow into a rainfall collection pipe. And the episode of the yellow runoff began.

Workers inspect water for human consumption daily at the storage reservoirs, and an alert technician noticed the water quality was going down the drain rapidly, a most unusual situation. Within a few days, the required dosage of chlorine climbed from the normally small and infrequent amount to levels reminiscent of a public swimming pool. Something was clearly amiss. The need for ever-increasing water treatment and a quick review of recent work on the overall plumbing systems led to the discovery that drain P(ee) in the men's restroom had been attached to a pipe R(ain). Public exposure of the situation would have been bad P.R., indeed. Staff immediately rectified the mistake, drained water from the contaminated reservoirs, and began the tedious job of refilling them with water trucked from wells near sea level.

Though this story of pernicious piping carries an unsavory flavor, the mistake compromised no one's health. The doses of chlorine saw to that. And although the thought of drinking human urine may be one of gut-wrenching disgust, urine of healthy people is, in fact, antiseptic. It will never replace good old potable water as the universal drink of choice, but it can be consumed without fear of instant or even lingering death.

As for psychological distaste, ask an astronaut about the virtues of urine. True, none of our manned space flights to date has lasted long enough to require reuse of human urine, but the day is coming when space travel of such duration will require astronauts to drink processed urine. In preparation, people have done so under tightly controlled conditions during experiments here on Earth. Those people are still alive and healthy. One may even be your neighbor.

As more reassurance that yellow runoff at the Volcano House did not threaten public health, consider the sources of much of the nation's drinking water: rivers and reservoirs containing raw and untreated liquid that would turn almost any stomach. Domestic livestock, wild quadrupeds, birds, and recreating humans use many of these sources of domestic water as open-air toilets. Filtering and chemical treatment save the day—and plenty of lives.

I have more recently made another association with yellow runoff, but one with savory, even pine-scented, connotations. As fate would have it, twenty-five years after the Volcano House escapade, Anne and I found ourselves living where seasonal yellow runoff is the rule. This more-recent Duffield house and its self-contained water system sit in the forested mountains of northern Arizona, where catchment of rainfall off a roof for domestic uses is common. Northern Arizona sits on one of those few continental tracts where the water table lies too deep for anyone but King Midas to afford the cost of drilling a water well.

The design of an off-the-roof water system in the high mountains of northern Arizona resembles the one in Volcano, except that the Arizona storage tank must be buried or insulated in some other way so it doesn't become a huge ice cube during winter. The collection of yellow-tinted water comes each year when the pine trees produce prodigious amounts of yellow pollen. The pollen coats the house roof before the next rainfall washes it into the water tank. Getting rid of this ocherous stuff is as easy as installing a filter along the pipe that carries water back into the house. Knowledge of where Arizona's yellow runoff comes from is a comfort, as well as an entertaining annual reminder of the great yellow runoff at Kīlauea.

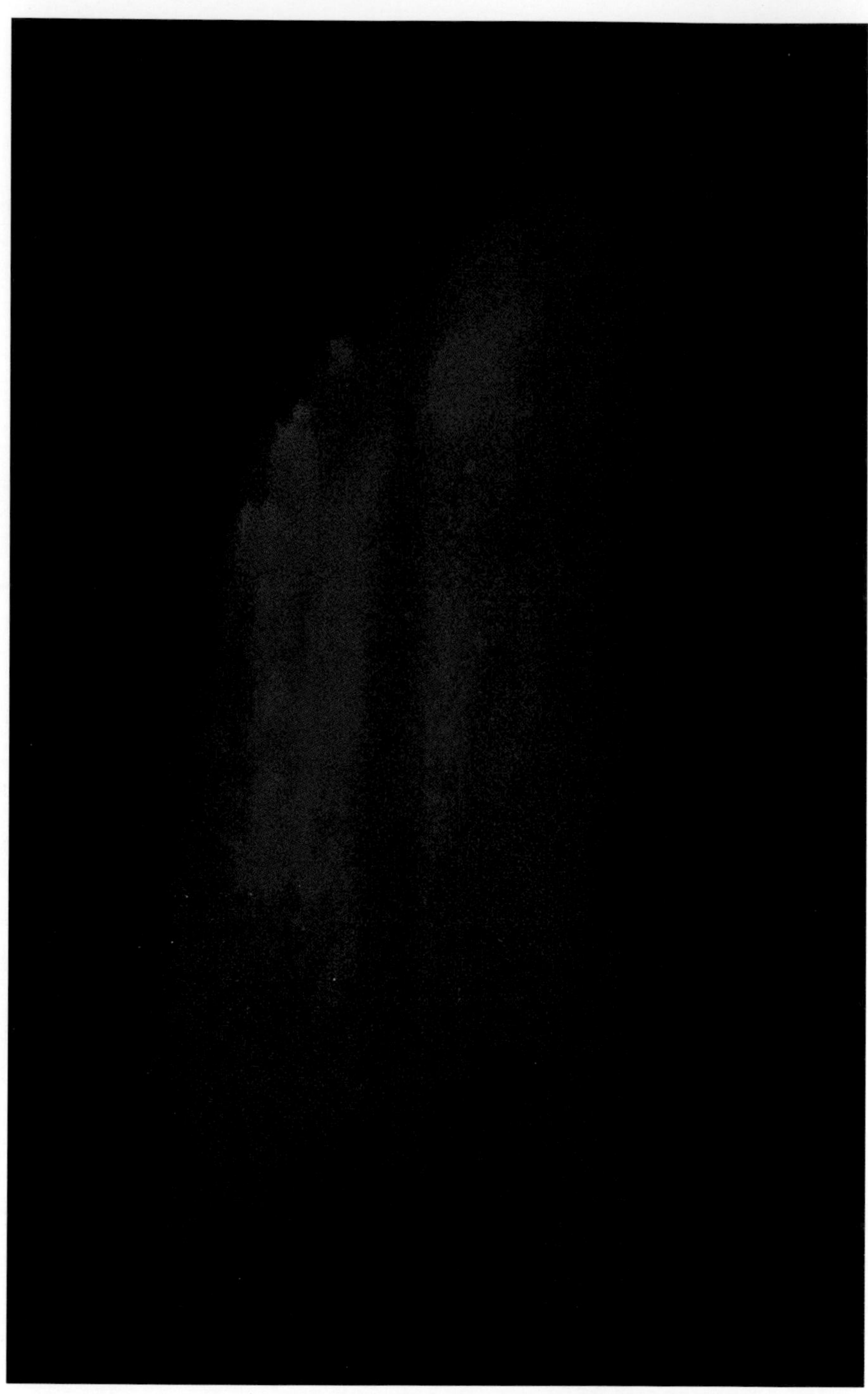

Mauna Ulu in magnificent nighttime eruption, 1969—U.S. Geological Survey photo

12

WATER BEDS AND MAGMA BEDS

In the 1960s and early 1970s, love, peace, long hair, and free everything was the order of the day. The waterbed became a popular and important part of that lifestyle. Used properly, a waterbed purportedly would enhance one's love life. Today, I don't know of a single friend or acquaintance who sleeps on a waterbed, and I don't think it's because we have become an old and stodgy generation. I think that conventional beds simply are superior sleep inducers, and let's face it, people of any age spend at lot more time asleep than at play on a bed.

As creatures of the Flower Child generation, Anne and I tried a waterbed for one night. We were passing through King City, California, on vacation, not long before we moved to Hawai'i. In this case, our curiosity got the better of our common sense. It only took a few moments between the sheets for us to realize our mistake, when the fun of sloshing around in a semicontrolled surfing adventure wore off. The water in our bed was way too cold for anyone to sleep on—at rest we shivered uncontrollably. We solved this problem, much to the befuddlement of the motel manager, by placing several extra blankets under us as insulation from the cold. We also turned the heater to its maximum setting, but thermal inertia kept the water in the bed uncomfortably cool through the night.

At HVO, I once spent an early morning on a magma bed with Don Swanson. The fluid in this bed was uncomfortably hot, and adding a layer of insulation was not a practical solution to the problem. Management in this case was Pele, the Hawaiian goddess of fire, and she was not about to intervene on behalf of two intruding geologists. Don and I created our own solution: a hasty retreat from the surface of the restless hot monster. The magma bed story starts with the Mauna Ulu eruption.

When I arrived in Hawai'i to begin my stint at HVO, an eruption was under way on Kīlauea's east rift zone. That eruption had begun three months earlier on fairly flat ground covered with rain forest, along a 2-mile reach of newly formed cracks and rifts. Within days, the rift zone spewing out magma focused its activity at about the center of a roughly equilateral triangle whose corners were the two preexisting craters, 'Ālo'i and 'Alae, and the cinder cone Pu'u Huluhulu. During the following twenty-nine months, sporadic eruption at this spot built a 250-foot-tall mound of lava, called a lava shield. The eruptions also fed some flows that extended across the south flank's grand staircase and into the Pacific Ocean, about 8 miles away. As the lava shield around the vent grew to become a notable figure of the landscape, the U.S. Geological Survey appropriately named it Mauna Ulu, Hawaiian for "growing mountain."

Before Mauna Ulu appeared on the landscape, the Chain of Craters Road continued across the south flank of Kīlauea to the south coast and then looped eastward near sea level to connect with a system of paved roads that gave access to the entire southeast part of the Big Island, including Hilo. Lava flows from early eruptions at Mauna Ulu, however, buried several miles of the road, greatly complicating travel between the summit of Kīlauea and the parts of the National Park along the south coast. With partial burial, the Chain of Craters Road simply led to a parking lot next to 'Ālo'i, beyond which the general public was excluded but HVO staff was expected.

Though stretched thin with leveling, geodimetering, and other studies related to deformation of Kīlauea Volcano, the HVO staff also maintained a frequent visual monitor of activity at Mauna Ulu. For most days, this

A nighttime eruption at Mauna Ulu feeds streams of lava that spill into ʻĀloʻi Crater.
—Wendell Duffield photo, U. S. Geological Survey

included a half-mile hike from the new abrupt end of the Chain of Craters Road, right up onto the Mauna Ulu shield.

Don and I often hiked to Mauna Ulu together. As our total number of round-trips grew substantially, but the landscape over which we hiked changed relatively little, at least on a daily basis, we became complacent about what was underfoot. This carelessness helped lead to the magma-bed incident.

One day on the hike to Mauna Ulu, as Don and I were walking and talking and perhaps not paying enough attention to our footing, we both suddenly felt the ground beneath us move in a soft and mushy way. This was not an earthquake. Simultaneously, and without a word spoken, we realized that we were literally on thin crust, Pele's magma-filled equivalent of a waterbed.

Apparently, overnight Mauna Ulu had erupted a piddling bit of lava, which had puddled in a low spot along our trail. By the time we arrived on the scene, a lava crust had formed over the puddle. But hidden beneath that skin lay a reservoir of still-molten rock. Though cooling and thickening with time, the crust was still so thin, probably only a few inches, that the weight of our bodies was pushing the crust down into the melt, just as a human body pushes the rubber bladder of a water bed into its liquid interior. We quickly backtracked to solid ground and waited for our heart rates to approach normal. I don't remember what happened next, but I never

again walked cavalierly onto very new-looking lava without first convincing myself that only solid rock lay underfoot.

In fact, we probably were never in imminent danger. Obviously, the crust, though thin, was sufficiently strong to hold our weight. Even if the crust had cracked, we may well have been able to walk safely away. The melt beneath the crust was thick, pasty, and viscous enough that we could have stepped from one crack-bounded piece to another before our weight pushed any one piece into the underlying hot ooze. Mind you, I wouldn't want to try this hopscotch dance, but I think it could be done. If the trapped melt was still frothy with gases, a different and more frantic dance might ensue.

The truly frightening thought is what might have happened if we had strolled onto the lava during what geologists call crustal overturn. Many direct observations at Kīlauea over the past several decades verify that a pond of lava may evolve through the following series of steps as it changes from hot and restless melt to solid rock.

> ***Step one:*** As soon as the pond forms, or even as melt is still being added, a thin black crust coats the surface. This happens as the melt, which solidifies at about 2,000 degrees Fahrenheit, quickly cools when it is exposed to the earth's atmosphere, which at Kīlauea is no more than 100 degrees Fahrenheit.
>
> ***Step two:*** With time and cooling, this crust slowly thickens. Simultaneously, gases (mainly water vapor, carbon dioxide, and sulfurous vapors) slowly and continuously escape from the underlying melt and collect under the crust.
>
> ***Step three:*** If enough of these gases collect, they buoyantly lift the crust until it breaks into pieces that tilt and sink into the underlying less-dense, frothy melt. This is called crustal overturn, because the process destroys the existing crust and new crust eventually takes its place.
>
> ***Step four:*** Depending upon the amount of gases remaining in the melt, steps one through three repeat, or step three ends with a stable crust that thickens until the entire pond of melt has solidified.

If the "crust" on your waterbed cracks and springs a leak, at the worst you will get soaked and have a wet mess to mop up. However, if the crust on your magma bed springs a leak and founders, you probably will have no mess to clean up. You will instead become completely incinerated toast. Perish the thought.

The exposed inside of this four-foot-tall tree mold shows the tree's shape and texture frozen into lava as it chilled against the trunk. —U.S. Geological Survey photo

13
TREE MOLDS AND LEG MOLDS

When a new lava flow spreads out across the rain forest at Kīlauea, interactions abound between the hot molten rock and the wet, green, inhabited landscape. Hawaiian lava is of such low viscosity, and therefore runny, that it typically spills into open fissures, pits, and other low spots, completely filling them if enough new lava is erupted. Hills and holes tend to disappear, or at least their sharp edges and vertical relief mute. While I was at HVO, lava that erupted at Mauna Ulu filled both the 500-foot-deep ʻAlae and the 200-foot-deep ʻĀloʻi pit craters. Hills and other places of high relief lose stature as lava encompasses and builds around them. In this manner, Mauna Ulu eruptions reduced nearby Puʻu Huluhulu's 300-foot height by several tens of feet. The net effect is similar to what happens when a continental glacier grows and flows, though the lava smoothes the preexisting landscape by adding new material to low ground rather than by scraping off high ground.

Moisture held in plants, soil, and near-surface rock follows one of two paths. When the amount of moisture and the rate of burial by the hot advancing lava are appropriately balanced, steam generated by the sudden imposition of a 2,000-degree-Fahrenheit environment simply hisses off into the atmosphere. Under different conditions, and I'm not at all sure what the exact differences are, steam and methane (from rotting vegetation) escape in the high-velocity bursts of localized explosions. Some colleagues and I discovered this explosive mode by accident while watching, at close range, the front of a flow advance through the forest. Needless to say, we avoided walking too closely along a forest-and-lava-flow interface thereafter.

As a flow advances, mobile creatures such as birds, bees, mongoose, feral pig, and feral goat flee. People also almost always have time to move far out of harm's way, commonly with all their easily movable possessions in tow. During the 1980s, faithful parishioners saved a Catholic church from a volcanic burial by mounting the still-usable building on wheels and hauling it several miles down the road in front of advancing lava. This is the first example I know of in which the church was literally taken to the people!

Molten lava flows burn and bury small and low-to-the-ground vegetation. This includes plants like ferns, grasses, pohā, and guava bushes. Typically, no physical trace of such small vegetation survives in the solidified lava, though a plant mold and a bit of charcoal might remain under appropriate conditions. Most or all of the plant matter burns and vaporizes, the produced gases mixing with and disappearing into the earth's atmosphere.

Lava's interactions with full-grown trees, however, are another story. First and foremost, lava instantaneously chills to solid rock an inch or so thick

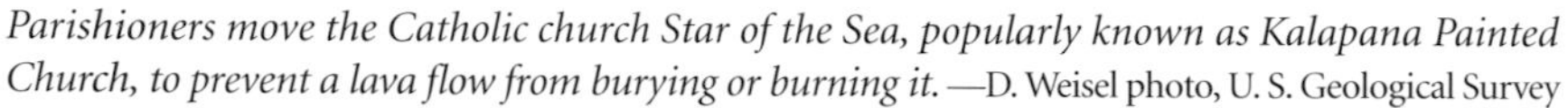

Parishioners move the Catholic church Star of the Sea, popularly known as Kalapana Painted Church, to prevent a lava flow from burying or burning it. —D. Weisel photo, U. S. Geological Survey

As a lava flow moves through a forest, it chills against and molds around tree trunks. The flow's intense heat severs the above-lava part of the tree, which bursts into flame as the flow carries it along. —Don Peterson photo, U. S. Geological Survey

around a tree trunk, thereby forming a tree mold. The above-lava part of a tree, though, has a very different history. Under normal circumstances, a full-grown tree is much taller than a flow is thick, perhaps 5 to 10 times taller. These woody sentinels stand briefly above the heat and turmoil but ultimately fall victim to the incinerator around their base. All encompassed trees burn off at the surface of an advancing lava flow. The fire-cut treetops fall on crusted but still-moving lava and head off downstream in a bouquet of flame. The visual effect is especially captivating and mesmerizing on a clear night, when flaming treetops litter an entire moving lava surface. We spent considerable time watching such scenes during the occasional gushes of great sheet-flood lavas from Mauna Ulu. Once burn-off is widespread, the landscape resembles a former forest clear-cut at the new ground level.

The rooted trunk left behind, with its tree-mold rock rind beneath the new lava surface, undergoes one of two transformations. If the atmosphere and its component of oxygen can circulate down along the lava rock surrounding a trunk, the wood burns just as it would in a stove or bonfire. The end result is a trunk-shaped hole in the solidified lava flow, with a bit of combustion ash in the bottom.

Without access to atmospheric oxygen, however, a lava-enveloped tree trunk undergoes nature's version of the high-school chemistry experiment called destructive distillation. In the experiment, the student puts a small piece of wood in a test tube and stoppers the tube so that, upon heating, vapor can escape but air cannot enter the tube. The student then heats the tube over the flame of a Bunsen burner and waits until he or she can no longer hear vapor hissing out the top. In the end, the tube contains a piece of charcoal, the carbon-rich residue of wood devoid of all volatile constituents.

Lava surrounding the trunk of a tree serves as both laboratory test tube and Bunsen burner. And at the end of Kīlauea Volcano's act of destructive distillation, the lava-enveloped tree trunk is nothing more than a large, cylindrical piece of charcoal. I've known people to collect such stuff and use it to barbecue hot dogs. Geologists enjoy barbecues, too, but they would cook with the charcoal of a tree mold only after recovering enough to determine the age of the tree, and thus the age of the lava flow, by the carbon-14 technique—unless they had witnessed the eruption and so already knew the flow's age.

Whether a tree trunk stranded in a lava flow ends up as a stick of charcoal or a small pile of ash, the shape of the trunk is preserved in the lava that chilled against it. That chilled lava molds against the tree, typically recording details of bark structure and texture. Entire "forests" of tree molds are scattered here and there on Kīlauea. Some stand as black sticks because most of the lava that flowed around them drained away as eruption waned. Others are simply cylindrical holes in the flow that spread around them. One tree-mold-stick forest on Kīlauea creates a landscape so unique and bizarre that it ranks as a state park.

Though no one would purposely run the destructive distillation experiment on a person, unplanned experience has demonstrated twice in the past twenty-five years that the human trunk acts much like a tree trunk when a lava flow envelops it. The first human guinea pig of this sort that I know of is Jeffrey, a technician at HVO during my time there. Jeffrey's unfortunate accident, thankfully, was not too unfortunate, because he is a living and fully functional human who will tell his own version of the experiment if appropriately prompted.

Jeffrey's unintended experiment happened when he was attempting to collect a sample of fresh lava from the edge of a flowing lava stream. We all had experience collecting samples of pasty, taffylike lava from active lava flows. Wearing noncombustible clothing, which protected our skin from the intense heat radiating from the lava, we walked up to the edge of a slow-moving flow. Then we dipped the sharp end of a geology hammer into the lava and extracted a gooey gob of melt, which transformed to solid rock during the tens of seconds it took to back away to a cooler place. Rock samples don't get much fresher or more pristine than those.

This tree mold stands above the lowered lava surface after an eruption stopped and the lava drained away. Maurice Sako for scale. —Wendell Duffield photo, U. S. Geological Survey

When Jeffrey collected his fated sample, the margin of the rock he stood on broke off and fell into the flowing lava. His right leg followed this rock ledge, sinking in about up to the knee before he could fall backwards and extract the limb to avoid additional harm. Fortunately, he had company who provided moral and physical support. A call on the two-way radio brought quick evacuation to the Hilo Hospital. He suffered some second- and third-degree burns on his leg, but the permanent damage was minor and cosmetic.

Jeffrey may have been the first documented demonstration of a human leg-mold episode interrupted early enough to avoid the creation of charcoal and to suffer only minor burns. Here is how that may have happened.

Molten basalt hardens to solid rock at about 2,000 degrees Fahrenheit. So a layer of solid rock quick-chills over the surface of a lava flow and around a tree trunk almost immediately when the lava contacts the atmosphere and the tree. Solidification takes much longer beneath that solid layer simply because the rock crust insulates the molten lava and greatly slows the transfer of heat and its attendant temperature drop. When Jeffrey's leg landed in the lava, the contact between his boot-and-pant-covered leg and the lava was akin to the contact between lava and a tree trunk. The immediate effect was to form a thin rock skin around his leg. This solid-rock mold insulated Jeffrey's leg from the 2,000-degree temperature that was just an inch or so away. By quickly withdrawing his leg and removing the lava mold, he minimized both the time and temperature of exposure, sparing himself from serious injury. But what a nerve-wracking experience.

The whole process was a bit like dipping an ice-cream cone into a large pot of molten chocolate. If the ice cream is left in, it will melt and mix with the tasty liquid sauce. However, if quickly removed, a rind of crisp chocolate molds around the cold interior, creating a chocolate-coated ice cream treat. Anyone who has enjoyed a chocolate-dip cone knows that about the first nibble at this treat often shatters the chocolate, which falls away in pieces that stain one's favorite white shirt. Jeffrey did not mind at all when his leg mold shattered quickly and fell to the ground—that lava mold was still pretty hot!

About fifteen years after Jeffrey's incident, a geologist named George suffered the two-legged version of Jeffrey's leg-mold experiment. Leg molds are not charming in any number, and certainly two episodes of leg-mold creation are enough. Tree molds in great profusion, though, make beautifully bizarre forests, and the more the merrier.

Jeffrey carefully approaches a sluggish pāhoehoe lava flow and breaks through the flow's thin black skin with his geology hammer. He pulls out a sticky sample that freezes to hard rock on his hammer within a minute or so.—Wendell Duffield photos, U. S. Geological Survey

Two-thousand-foot-tall eruption at Mauna Ulu feeds streams of lava that spill into ʻĀloʻi Crater.
—Wendell Duffield photo, U. S. Geological Survey

14

A SUFFOCATING EXPERIENCE

Sooner or later, we all learn that the road of life contains a plethora of hillocks and potholes, the figurative and sometimes literal causes of life's ups and downs. We cope by circumnavigating these obstacles when possible or by slowing down enough to make a rough ride tolerable. With a splash of optimism and a positive attitude, most of us can manage even a fairly bumpy ride, especially when we can see the ups and downs. The invisible obstacles, though, can cause great immediate grief and lingering anxiety. When the enemy is silent, odorless, and absent to the touch, as well as invisible, one can feel totally helpless and susceptible to paranoia. As if the magma-bed incident had not wrought punishment enough on two geologists intruding on Pele's ground, this goddess placed an invisible obstacle before Don and me during another of our routine checks on Mauna Ulu.

The day was calm and dry, imminently suitable for a quick visit to Mauna Ulu to see if anything volcanic had transpired during the previous twenty-four hours. In hindsight, though, this particular day was somewhat unique in that the usual trade winds were not blowing. The air was eerily and exceptionally still.

Our magma-bed experience fresh in our minds, we were appropriately cautious to avoid repeating such hotfoot dancing. But we were not prepared for the vaporous danger that lay hidden along our path on this otherwise tranquil day. The superficial circumstances were much like those on the magma-bed day. We were walking together, chatting about volcano business and sensing nothing amiss. Then, just as with the magma-bed realization, we stared at each other in simultaneous recognition that something was terribly wrong. We could not breathe. And in magma-bed fashion, our solution was simultaneous retreat along the trail as quickly as possible, knowing that breathing had been the usual involuntary routine just a few steps back.

Lungs once again awash in that elixir of oxygen-rich air, we arrived, after just a few moments thought, at a very logical explanation for our problem. Whereas a lava flow had ponded along a low section of the trail in the magma-bed incident, this time carbon dioxide had accumulated in a low pocket, at the expense of the normal mix of atmospheric gases, including oxygen. At the time, we had no way to verify our theory, no instrument to measure the concentration of carbon dioxide along that section of the trail. Even if an instrument had been available, we probably would not have wanted to reenter the suffocation zone to collect a gas sample for analysis. But no other reasonable explanation occurred to us then or after we reflected on what had happened.

Volcanoes do emit carbon dioxide, among other gases. These gases are dissolved in the magma when it begins to rise from the earth's mantle, where molten rock originates. The gases start to come out of solution near the earth's surface, where confining pressure is too low to keep them dissolved in the magma. The rapid and vigorous escape of gases is what drives lava fountains high into the air. As in uncapping a bottle of carbonated soft drink, the magmatic gases are essentially "uncapped" from solution in the magma by rising to the earth's surface. And this analogy is very appropriate in our case, since the pertinent gas in both situations is carbon dioxide.

The most abundant gas dissolved in magma is water. When released from magma and mixed into the atmosphere, this stuff is water vapor, which is safe to breathe. In the tropics where the relative humidity of the atmosphere is almost always high, we breathe plenty of water vapor regularly.

Another fairly common suite of gases in magma consists of sulfur-bearing compounds, most notably sulfur dioxide (SO_2) and di-hydrogen sulfide (H_2S). The presence of sulfur-bearing gases in volcanic areas is usually obvious from their rotten-egg odor and throat-irritation properties, which drive both tourist and scientist away quickly. There is no hidden danger here.

Carbon dioxide, though, is an insidiously dangerous magmatic gas. It is odorless and colorless, and most significantly, it is more dense than the mix of gases that make up the earth's atmosphere. If you ever had an introductory course in chemistry, your instructor may have demonstrated this density difference by pouring carbon dioxide from a glass beaker over a burning candle. Though totally invisible, the student "sees" the poured carbon dioxide push against the flame and then extinguish it as the gas displaces all oxygen from around the candle.

In practical terms for living in the great outdoors, this density difference means that in an adequately windless setting and where a source of the heavy gas exists, carbon dioxide will flow into and fill low places simply under the pull of gravity. That fateful day in 1969, carbon dioxide spilled out of the Mauna Ulu vent, and the windless conditions permitted the gas to collect in a pool along the trail that Don and I walked. Any living thing that requires oxygen will suffocate in such pools of carbon dioxide. We quickly recognized our problem and made a hasty retreat to a place of normal atmosphere. In other situations, though, people and other animals have not been so fortunate.

In 1986 at Nyos Volcano in Cameroon, Africa, a giant nighttime burp of carbon dioxide suffocated about 1,700 people and 3,000 of their cattle. Nyos had last erupted about 400 years earlier, creating a crater 1 mile wide and 1,000 feet deep, which quickly filled with water. Over time, the water of Lake Nyos became saturated with carbon dioxide that continued to rise from the deep magma source, through the roots of the volcano's conduit, and into stagnant cold water in the bottom of the lake. In effect, the lake water, and especially the deepest and coldest water, became a highly carbonated beverage whose cap was removed when something stirred the lake enough to upset equilibrium. Once "uncapped," rapid effervescence sent an invisible river of carbon dioxide flowing down an adjacent stream valley, suffocating people and cattle up to 6 miles distant. This was a tragic real-life field illustration of the candle-snuffing laboratory experiment. While somewhat unusual circumstances led to such a voluminous and instantaneous outpouring of volcanic carbon dioxide, the event serves as a reminder that even sleeping volcanoes can be killers. Fortunately, no such carbonated crater lake exists at Kīlauea.

A pessimistic and fatalistic soul might conclude that Pele paved the path to Mauna Ulu with perilous pitfalls. True, dangers do lurk along the trail, and Don and I seemed to experience them often enough. But we escaped all harm. Still, the odd pool of carbon dioxide that Don and I encountered is a danger worth keeping in mind if you should go volcano watching on a windless day.

15

HOW FAST CAN YOU RUN?

This tale, one more Mauna Ulu story, focuses on a geologist's complacency, a complacency that is difficult to condemn but worthy of a chastising note. It illustrates that we should never take for granted even such relatively benign volcanoes as those in Hawai'i.

A popular pastime for several of us on the HVO staff was to watch Mauna Ulu's lava fountains from the top of nearby Pu'u Huluhulu. This prehistoric cinder cone was the perfect platform for viewing Mauna Ulu. When Mauna Ulu first began to grow, Huluhulu stood nearly 300 feet higher and about 2,000 feet to the side of where the new volcano sprouted. And even after all the eruption phases ended, Huluhulu was only a bit lower than its new neighbor was. Moreover, the usual trade winds blew volcanic fumes and the occasional errant piece of falling cinder away from Huluhulu, keeping us out of harm's way.

We watched Mauna Ulu both for business and for pleasure. As scientists, we gathered information that would help us understand volcano behavior. But mainly what kept us coming back was the sheer pleasure we felt while witnessing, first-hand and live, the beauty, grandeur, and power of the earth as she spilled her molten guts out onto the landscape. Very few people have such an opportunity. Those who do never forget the sights, sounds, smells, and feeling of the experience.

Our favorite eruption scene was lava fountaining high enough to stir our sense of awe but not so high as to seem dangerous from our vantage point on Pu'u Huluhulu. Both day and night offered spectacular viewing. Daytime viewing brought a wide array of sights and sounds. Nighttime presented the simple and stark two-tone spectacle of a bright orange, swordlike shaft of lava stabbing upward into the black of night.

Most people who have heard lava fountaining describe the sound as like that of a jet aircraft taking off at full throttle. The eruption sound reflects the upward jetting and thrusting of clots of melt as rapidly expanding magmatic gases drive them skyward. Though in my experience the decibel level never drowned out neighborly conversation, the constant rumble of fountaining lava resembles the booming of a powerful electronic amplifier driving huge bass-range speakers.

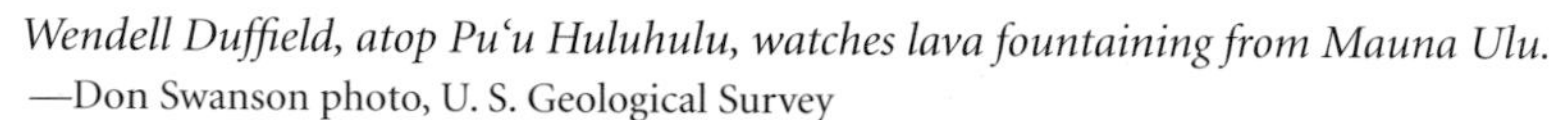

Wendell Duffield, atop Pu'u Huluhulu, watches lava fountaining from Mauna Ulu.
—Don Swanson photo, U. S. Geological Survey

The physical feel of an erupting fountain, fortunately for the viewer, is an indirect sensation, transmitted to the human body as a low-amplitude, steady shaking of the ground. City dwellers might compare the sensation to vibrations they feel along a subway track as a train passes. Farm boys like me might compare the feeling to vibrations emanating from a thresher, combine, or other large farm machine in operation. We volcano viewers could easily feel the eruption's vibration at Huluhulu if we simply stood very still.

Seismometers up to several miles from the source can record this fountain-caused ground vibration. Seismologists give this earthquake signature the name harmonic tremor, in recognition that the seismic record shows an up-and-down wave form of fairly constant amplitude that repeats itself about every second or so.

One way to picture the origin of harmonic tremor is to compare it to a grand concert-hall organ. With this musical instrument, air surging through a pipe creates a pleasant-sounding vibration in the frequency range audible to the human ear. The goddess Pele plays comparable "music" in the form of a ground vibration when she sends magma flowing through a pipelike conduit of rock. I find that both forms of vibration can excite yet another form of shaking, a pleasurable chill that travels up and down my spine.

One fine day on Pu'u Huluhulu, Hans, a geologist visiting from Germany, and I were absorbing the sight, sound, and feel of lava fountains on Mauna Ulu. The fountain heights peaked at a few hundred feet, a bit above our vantage point, and in its normal configuration, the lava "gun" pointed straight up. Thus, fallback splattered closely around the vent itself. Clots of melt that solidified to solid rock during flight built up a cinder and spatter cone, while the still-liquid clots joined lava flows oozing away from the base of the cone.

Suddenly and with no recognizable warning, Hans and I found ourselves squarely in the fallout zone. Inch-wide pieces of hot rock and still-molten clots were targeting Pu'u Huluhulu, and Hans and I stood perilously close to the bull's-eye. Without hesitation or discussion—for even the most loquacious and impractical of academicians will waste no time trying to explain what is happening while "fire" is falling all around him—we ran directly away from the offending fountain.

Since the established foot trail on Pu'u Huluhulu included a series of switchbacks across the slope facing Mauna Ulu, we crashed through the tangle of rain forest in the opposite direction. We were wearing metal hard-hats like those used at construction sites, and the clots of hot, pasty cinder ricocheted off our helmets with unmistakable pings, clunks, splats, and thwacks. Both Hans and I garnered a few burn holes in clothing and minor skin burns where clots penetrated cloth. I was wearing my Oʻahu State Prison shirt, and it was never the same after this shower of fire. After just a few minutes and couple hundred yards of crashing cross-country through the forest, we were out of the danger zone. *Whew* and double *whew*!

A few minutes later, the lava fountain redirected its brilliant shower to the usual vertical orientation, making Huluhulu a safe viewing platform again. However, Hans and I were not even slightly tempted to climb back up that hill.

When fountaining ended later in the day and we could safely reconnoiter the effects of eruption, we found the side of Puʻu Huluhulu that faced Mauna Ulu clad with a new coat of lava lacquer up to several feet thick. This new layer of fallout had been so hot and pasty when deposited that parts of it had slid and oozed several feet downslope in viscid sheets. The other side of Puʻu Huluhulu—the side of our escape route—still wore its rain-forest vegetation, damaged little by scattered clots of falling cinder.

Fallout from a lava fountain drapes the side of Mauna Ulu that faces Puʻu Huluhuluu. The fallout was so hot that it pulled apart along the dark, jagged breaks and flowed partway down the hill. —Wendell Duffield photo, U. S. Geological Survey

So why did the lava fountain on Mauna Ulu briefly spray Pu‘u Huluhulu? Almost certainly the vent hole in the ground from which lava squirted suddenly changed shape. Most commonly, a fountain shoots vertically, and geologists interpret the erupting conduit that feeds such a fountain as a vertically oriented pipe-shaped passageway. However, on that fateful day for Hans and me, an unplanned and unannounced obstruction partly blocked the pipe, temporarily diverting the lava stream sideways. Very likely, a large piece of rock fell in from the wall of the crater around the pipe. It served the same function as your thumb over the lip of a garden hose when you want to divert the water spray in a direction other than straight out.

Was Pele's thumb at work over the vent that day, reminding us that we study Kīlauea at her whim? As a scientist, I should answer that question with a resounding "no!" But volcanoes will always carry an element of mystery for me—so I guess I don't know.

With a dollop of imagination, the viewer of this lava flow near Mauna Ulu can see Pele shielding her face with her right hand as her left hand wields a long clublike object.
—Wendell Duffield photo, U.S. Geological Survey

When Pele weeps, she sheds tears of molten lava, which solidify into a variety of bizarre shapes as they fall to earth. —Wendell Duffield photo, U.S. Geological Survey

16

LIVING OFF THE LAND

In spite of all of the various HVO volcano projects, Anne and I occasionally found time for pure unadulterated recreation. Because the entire HVO staff was expected to be on call twenty-four hours a day, seven days a week, our fun time never took us far or long from Kīlauea. But it doesn't take much imagination to find pleasure in one's own tropical backyard. If the popular pidgin phrase "lucky you live Hawai'i" is to carry any substantive meaning, one should sink his teeth into and thoroughly taste the bounty of the tropics.

Anne and I have always viewed gardening as therapeutic recreation, so we tried raising a conventional vegetable garden at Mingo's house in Volcano. We had fair success in producing mature plants. However, the fruits of our labors seemed to disappear into the tummies of various marauding insects and other critters within hours of ripening. Unwilling to use chemicals that might have salvaged some produce for us, we gave up on home-grown food. Instead we harvested quite a bit of the wild-and-free variety.

A popular conception of life in the tropics includes dining from a cornucopia of fresh and free food, there for the taking. In my experience, this notion is partly correct. I never found a well-balanced diet of free food on the slopes of Kīlauea, and an increasingly dense human population has reduced the variety and availability of such goodies since I lived there. But you can still find the treats if you know what to look for and where.

While working near sea level on the trade wind (wet) side of the volcano, we commonly encountered banana, coconut, guava, mango, and papaya in the wild. Most guavas are yellow fruit, but the tastiest and rarest variety is called strawberry guava, which has both the color and flavor of strawberries. Mangoes were especially abundant and available, and in high season they even became a traffic hazard. Fruit-laden canopies of huge mango trees overhang many of the roads on Kīlauea, and when the overripe fruit splats to the pavement, it converts the road surface to slime. We supposed that someone had once cultivated, and later abandoned, many of the fruit-bearing plants. Whatever their history, we helped ourselves to what was ripe so long as we detected no apparent ownership.

At the much cooler elevations around the summit of Kīlauea, where we did most of our fieldwork, wild fruit was far less abundant. We did find the occasional guava bush and small tree there. With some searching, we could find pohā in the rain forest. Pohā, known as ground cherry on the mainland, looks like a small tomato. This walnut-sized, soft fruit tastes good right off the plant, and it also makes delicious jam. Breakfast at the Volcano House featured pohā jam when I lived in Hawaiʻi, and perhaps it still does. At some time, presumably after Europeans came to Kīlauea, someone planted an avocado tree near the caldera. The tree's exact location was (and still is?) a secret shared only with those deemed trustworthy to harvest for themselves only a small share of the season's crop.

When it came to wild foraging, though, my true love affair was with ʻōhelo berries. ʻŌhelo berries grow on scraggly, woody bushes that rarely get taller than a foot or two. These plants seem to thrive only around Kīlauea Caldera, as though some unique contract of coexistence binds plant to volcano. Apparently, ʻōhelo berries find a chilly and windy environment frequently bathed in smelly sulfurous volcanic fumes the ideal setting for growth. Plants are most abundant and robust just north and south of the wall of Kīlauea Caldera, within a zone of transition from the wet, trade-wind side on the east to the Kaʻū Desert rain-shadow side on the west.

Native Hawaiian culture considers ʻōhelo berries sacred to the goddess Pele, and Pele's home is at Kīlauea. So while a few people report that ʻōhelo berries also grow on Haleakalā Volcano, on the neighboring island of Maui, these berries do not grow under the conditions of an active volcano. Perhaps they are pretenders or merely leftovers from when Haleakalā was Pele's home. I believe Kīlauea Volcano is home to the one and only true ʻōhelo-berry belt.

ʻŌhelo berries, ready for picking and consumption —Jack Jeffrey photo

A ripe ʻōhelo berry is about the size of a green pea and pale yellow to orange to eye-catching red. Berries are edible, though nearly flavorless, right off the bush. They are equally flavorless after mellowing a while in the refrigerator. Most connoisseurs of this berry make them into pies, apparently under the dictum that enough pectin and sugar transforms almost anything into yummy dessert. Anne made, and we ate, many ʻōhelo-berry pies. They were filling, but less tasty than an apple pie.

Before you travel to Kīlauea to sample this rare fruit, you should be aware that to eat ʻōhelo berries is to also eat the small white worm that appreciates this berry as much as or more than any human does. The odd worm-free berry exists, if only briefly, but you probably will never be able to harvest enough berries for pie unless you are willing to eat many worms. For those feeling squeamish about now, I'll add that these wiggly products of barely digested berries seem to carry no taste of their own. They really are what they eat.

Anne's and my favorite form of the fruit was ʻōhelo-berry wine. We discovered that fermentation effectively separated the worm from the berry's juice. The effervescence of fermentation carries the worms and any other solid ingredients to the top of the active must, and we disposed of these solids in one way or another in the wine-making process before bottling.

Professional vintners claim that a conventional red wine—that is, one made from grapes—derives much of its flavor and aroma from fermentation in the presence of the grape skins. Generations of experience have proven this claim to be fact. From our experience, we can say the same for fermentation in the presence of ʻōhelo-berry skin (and worm carcasses)—unfortunately, the skins, and the wine, are about as flavorless as the raw berry. The final product, though, is delightfully alcoholic.

We crushed berries, fermented their juice, and bottled the final product at Mingo's house in Volcano. After aging the wine for a few hours to several days, our HVO friends and we consumed some of it in pure and mildly chilled form. More often, though, we blended it with a rich white grape wine to add flavor and interest. Whenever I drank ʻōhelo-berry wine—tasty or not—I felt I was somehow bonding with Kīlauea Volcano and the goddess Pele, drinking the wine of the plant whose true home is Kīlauea Caldera.

17

CAMPING AT HALAPĒ

To escape the twenty-four-hours-a-day fireman's life at HVO, Anne and I would go camping. We were away from a telephone yet reachable by vehicle and/or foot if the need should arise. Most of our camping ventures took us to beaches on the rain-shadow Kona side of the Big Island. However, the island is so geologically young that the waves have not yet beat the coast's hard lava flows into soft sandy beaches, so there are precious few decent swimming beaches. Even a lovely Hawaiian beach can get boring after repeated visits. So one fine long weekend, we set off with HVO friends Dallas and Beverly to camp at a place called Halapē.

Halapē is in Hawai'i Volcanoes National Park, along the south coast of Kīlauea. This destination may not seem like a real getaway from the HVO job until you realize how isolated and even forbidding the place is. Our access was by foot only, with two options: a 10-mile walk along the south coast over rough lava flows, some from Mauna Ulu and, thus, only weeks old; or a 4-mile hike across the grand staircase of the south flank, losing about 2,400 feet of elevation in the process. Though the return hike would be steeply uphill, we chose the shorter route, mainly because we would be carrying in four days' worth of provisions.

From the end of a jeep road on the 'Āinahou Ranch, then a small privately owned piece of land surrounded by the national park, we headed down and south to Halapē early one morning. Within the first half-mile, we dropped nearly 1,000 feet as the trail descended over the first of the grand-staircase risers. This steep drop, called Poliokeawe Pali, is the topographic expression of the headwall of a huge landslide block, a piece of Kīlauea that broke free and slid partway into the ocean earlier in the volcano's history.

Next came a mile and a half of gentle downhill until we once again encountered a steep headwall, Hōlei Pali, only about 300 feet high. The final pull into Halapē took us down a lava ramp tilted to the southwest, a surface of densely fissured and broken ground that spoke of a tortured and violent volcanic history. The view to the horizon both west and east was one of complexly anastomosing pali and broken intervening ground, widespread and visually compelling evidence that this entire part of Kīlauea has repeatedly fractured and fallen apart in the volcano's recent past.

The only pristine and coherent lavas in our field of view lay to the east, where we could see a thin ribbon of flows that cascaded over and down the grand staircase from Mauna Ulu. Only a few months old, these had not yet experienced massive landslides like those that shape the grand staircase of the south flank.

Though we had no inkling at the time, even that new Mauna Ulu lava drapery across the grand staircase would soon break, during an earthquake and landslide event that entirely reshaped our campsite destination.

We reached Halapē at about noon, hot and sweaty but ready to set up the trappings of housekeeping known as camp. The usual wide, soft and sandy Hawaiian beach didn't exist here, so we made camp on rocky lava flows and on a patch of coarse sand and boulders that storm waves had tossed up. A small grove of several dozen coconut palm trees, a hundred feet back from the ocean, provided a shady refuge from the hot tropical sun. These trees provided us with free coconuts, too. A small shed sheathed in corrugated metal offered additional shelter from the sun, but the walls of the shed also blocked the cooling beach breeze, canceling the benefit of shade.

A family or two of mongooses viewed us as intruders on their homes in the coconut grove and beneath the shed. During our entire stay, they announced their displeasure of our presence with frequent attacks of hissing and screeching from just a few feet away. They also proved very adept at getting into our larder, in spite of our efforts to thwart their thievery. I fear that these impertinent and aggressive creatures will be a true hazard to people if rabies ever finds its way into Hawai'i. They are aggressive even in the undiseased state.

The waterfront itself was an abrupt contact between crashing waves and bedrock lava flows. We would enjoy no casual swimming or snorkeling here. About 200 yards offshore, breaking waves constantly battered a several-acre, 20-foot-tall island of naked lava flows called Ke'a'oi Islet. Despite the strong currents, Dallas and Beverly managed to swim to the island and explore that bleak piece of real estate. About the same distance inland, just behind our camp, broken and fissured lava flows, some with cracks gaping several feet wide, stood as clear evidence of the violent geologic events that shape the entire south flank.

Another thousand feet behind and a bit west of camp, a nearly vertical pali rises 1,000 feet to a prominence called Pu'u Kapukapu, meaning "forbidden (or sacred) hill" in Hawaiian. The hill stands as a lone peak along the crest of the pali and seems to look down, perhaps disapprovingly, on Halapē and our camp. One evening, Dallas thought he heard ominous, low guttural voices coming from Pu'u Kapukapu, but we later determined they were the grunts and snorts of grazing feral goats. Even a moderately inactive imagination might read an ominous message into our campsite setting.

But we ignored tall hills with foreboding names and long shadows, and we learned to coexist with the mongoose families. With no decent swimming beach, we passed our time reading books, talking, playing board games, and working on suntans, which were impossible to capture in our cloudy home of Volcano. Our time away from HVO passed quickly, too quickly. On the third day after arrival, we hiked back to our vehicle and drove home.

Since hindsight is almost always perfect, the skeptical reader may dismiss the rest of this chapter as writer-padding claptrap. Still, a sufficiently negative vibe surrounds Halapē that a sensitive person visiting there might easily adsorb a portent of bad things to come. The combined effect of a broken and tortured landscape, a dangerously energetic waterfront, a small sterile island, Pu'u Kapukapu overlooking camp, and even those aggressively feisty mongooses was not the product of a day at the amusement park. Halapē seemed a focus for many negatives.

Four years after our visit, a locally generated and very powerful (magnitude 7.2) earthquake violently shook the place. The temblor reactivated landslide blocks, which slipped farther toward and into the ocean, moving an additional 24 feet across and 11 feet down the grand staircase. At Halapē itself, the ground subsided about 10 feet. Most of Ke'a'oi Islet disappeared beneath sea level. In hindsight, it seems obvious that this island was nothing more than the barely emergent shoulder of a landslide block, poised to disappear into the ocean incrementally with each shake and slip.

Our campsite in the coconut grove dropped below sea level. Salt water flooded the grove and eventually killed the trees.

A troop of Boy Scouts and other campers were at Halapē when the earthquake struck. An earthquake-and-landslide-generated tsunami, or tidal wave, swept through the grove, killing the scoutmaster and a fisherman. Mongooses likely perished, too. When Dallas, Beverly, Anne, and I read about the event from our homes on the mainland, we felt that Pu'u Kapukapu, the sacred and forbidden mountain, had spoken to the intruding visitors. The Pu'u Kapukapu name now seems even more appropriate.

Halapē area before and after the earthquake and tsunami of 1975. The distinct linear color change extending left (westward) from Pu'u Kapukapu (top photograph) *reflects a change of vegetation across a fence.* —Don Swanson photos, U.S. Geological Survey

When lava flows from Mauna Ulu declared war on the Chain of Craters Road, this work of humans was defenseless. —U.S. Geological Survey photo

18 SWORDS INTO PLOWSHARES, SPEARS INTO PRUNING HOOKS

The closest that combat came to American soil during World War II was in Hawai'i at Pearl Harbor. Hawai'i was not yet a state but was a part of the United States of America in just about every other meaningful way. Geography alone exposed the Hawaiian Islands to possible aggression from the enemy. Not surprisingly, then, geography and the attack on Pearl Harbor produced a level of local and national concern that resulted in considerable military training and maneuvers in Hawai'i during the war years. The Big Island was one Hawaiian place of much mock warfare, and people still find military trash from these training and readiness activities on the slopes of Kīlauea. I became acutely aware of this refuse while mapping geological structures in the Koa'e fault system.

The Koa'e fault system is the 7-mile-long and 2-mile-wide central part of a structural boundary zone between Kīlauea's south flank and the summit area. The east and west ends of the Koa'e imperceptibly merge with the upper parts of the east and southwest rift zones, just below the rifts' intersections with the caldera. Together, the Koa'e and the lower parts of the rift zones define a physical detachment of the south flank from the rest of the volcano. The Koa'e shows most of the scars—the faults, rifts, fissures, cracks, and the like—of the geologic battle that has produced this detachment.

Finding the debris of military maneuvers in the Koa'e is fitting, for this fault system is a geologic battleground extraordinaire. Geologic forces have overwhelmed what was originally a gently seaward-sloping stack of thin basalt lava flows and turned it into a wounded, warped, bent, broken, fissured, and otherwise rent landscape of raised and lowered blocks of ground.

Most of the faults and fissures in the Koa'e fault system are the tapered distal extensions of cracks that originated near eruptive vents in the east rift zone and then propagated into the Koa'e. Cracks may also propagate into the Koa'e from the southwest rift zone, though the evidence is equivocal. Almost devoid of the eruptive vents themselves, the Koa'e fault system seldom floods with new lava and so provides a window into the probable subsurface structure of the rifts. The surface appearance of nearby parts of both the east and southwest rift zones might closely resemble that of the Koa'e fault system except that eruptions in the rift zones tend to bury fissures and faults about as soon as they form there. A newly opened eruptive fissure commonly totally inundates itself with its eruptive products, applying a kind of lava-flow bandage over the original gaping-fissure wound.

One fascinating and geologically useful trait of the chaotic Koa'e is that much of this fragmented terrain can be fairly accurately reconstructed visually and graphically into its original unbroken shape. For example, the irregular outlines of opposing walls of many gaping fissures fit back together like pieces of a jigsaw puzzle. Similarly, the lava on top of an uplifted block in some instances matches the lava on top of the adjacent dropped block. Geologists find this kind of mental and graphical reconstruction immeasurably useful in helping them understand the long-term structural history of the volcano. If only we could so easily mend human-war battlefields, and repair their damage, and somehow put them to constructive use.

My mapping in the Koa'e required endless on-foot traverses, miles and miles of walking to chart the details of geological disruptions of the ground. My extensive coverage of the area turned up the trappings of human as well as geological warfare.

Hairline extensional fractures, such as those near the 35-millimeter shell (inset), *grow incrementally into the major sinuous zones of broken and faulted ground that comprise the Koaʻe fault system.* —Richard Fiske main photo, Wendell Duffield inset photo, U.S. Geological Survey

I commonly found armor-piercing shells in my ramblings. These came in two sizes, one 35 millimeters in basal diameter and 110 millimeters tall and the other 75 millimeters by 210 millimeters. Universally, geologists seem to be collectors of no matter what, so I duly carried this trash back to my office. Solid steel projectiles that taper to a point, these shells are heavy. Over the years, and even today, they have served me well as paperweights and bookends.

The most abundant military trash I encountered, in number of pieces not weight, was pea-size spheres of lead. Such lead balls, or pellets, are spread widely though unevenly across the entire Koaʻe area. As a farm boy trained in duck and pheasant shooting, I thought these balls looked like unreasonably oversized buckshot. Even if only one of these gigantic pellets was to strike a pheasant, precious little of the bird would remain in a form worth eating. I learned later that these pellets are indeed a kind of buckshot, the kind intended for humans, under the euphemistic moniker of antipersonnel ammunition. I also later found and collected an unexploded shell that I probably would have walked right past had I known what potential fury it contained. The still-armed version was a thin-walled steel cylinder open at one end, 75 millimeters in diameter and 225 millimeters tall. The open end exposed lead pellets that a gummy substance held together. I assumed pellets filled the host cylinder to its bottom. How wrong I was!

I took the shell back to the observatory for John, our WWII-vintage machinist, to inspect. He was instantly horrified. He took the shell gently and submersed it in a tub of water, where it sat for days. The adhesive around the pellets released their hold during the soaking, and when John felt it was safe to inspect what was left, he showed me a metal plate positioned about 35 millimeters above the bottom of the otherwise hollow cylinder. Pellets filled the space above the plate; from the plate down was gunpowder, whose unfinished job was to explode and distribute the pellets in shotgun fashion. A small hole in this plate had taken in water during the long soaking and, thus, effectively defused the powder.

In the end, I had an empty steel cylinder, which now safely sits on my desk as a pencil holder. However, this example of swords into plowshares could have ended sadly because of my ignorance, stupidity, and catlike curiosity. The splayed remnants of another antipersonnel shell, one that apparently had exploded properly to distribute its antipersonnel buckshot, adorns my home as a multifingered "modern sculpture" mounted in a lovely oak block. Though not as utilitarian as a plowshare, pruning hook, paperweight, or pencil holder, this piece of art serves to remind me of the peaceful uses for military trash.

In the late 1950s, about a decade before my stint at HVO, certain military-weapon trash at Kīlauea made a creative and innovative transformation to geologic use. The problem Scientist-in-Charge Jerry Eaton faced was how to accurately measure small amounts of ground tilt to monitor the state of inflation or deflation of the magma reservoir beneath Kīlauea Caldera. Japanese scientists and technicians had already developed a workable technique for measuring tilt at their volcanoes, but their apparatus was not available for export. Scientists at HVO needed to somehow produce something similar to the Japanese system, for use at Kīlauea. The solution came in the form of artillery-shell casings machined into precisely graduated,

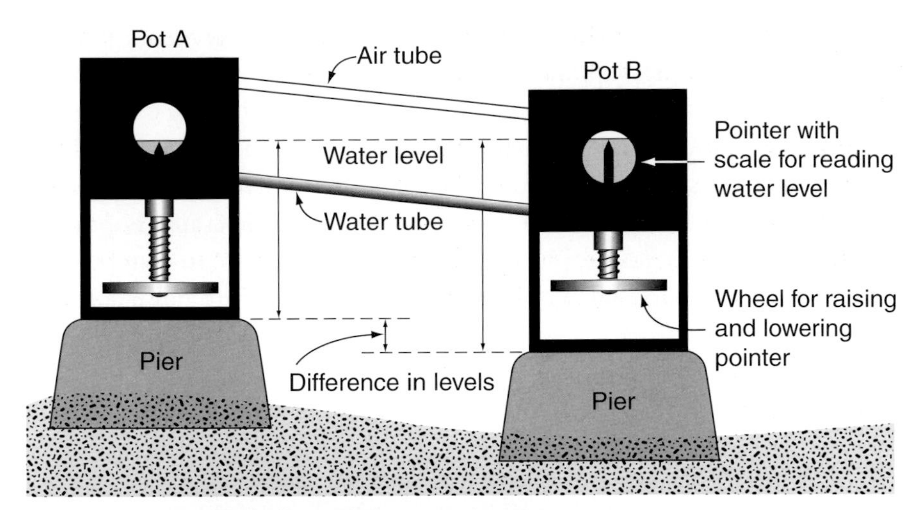

Willie Kinoshita reads the water level in a tilt pot to determine changes in the inclination of the volcano's surface since previous measurements. Jerry Eaton designed this system of "wet tilting" and had the tilt pots fabricated in the HVO machine shop from locally available World War II artillery-shell casings. —Richard Fiske photo, U.S. Geological Survey

interconnected water pots. The formal name that graces this instrument is a "portable water-tube tiltmeter."

Eaton based the design of this tiltmeter on the observation that water flows downhill. Thus, if the difference in elevation between two interconnected water pots changes, water will flow from the now-higher pot to the now-lower pot until the elevations of the water surfaces in each pot are equal. With three interconnected pots and repeat readings, Eaton could determine the precise direction, as well as amount of tilt.

The challenge was to make a device that could measure the change in water level in each pot accurately enough to prove useful in monitoring the behavior of the volcano. The machine shop at HVO had the equipment to precisely modify spent brass artillery-shell casings, which were readily available on the Big Island, into the water pots. From garden-variety

110-millimeter artillery-shell casings, the staff both designed and built a system that could measure tilt as small as a fraction of a part per million. That degree of tilt is the equivalent of tilt in a one-mile-long carpenter's level that has a thin coin under one end. So, instead of the military possibly reusing the shells as weapons to create explosion craters, geologists used the artillery shells to monitor activity around a volcanic crater, hoping to help avoid or at least minimize harm to people and damage to their possessions. May we see more such examples of swords made into plowshares!

19 DISHARMONIOUS TREMOR AND PREMATURE ERUPTION

If the scientist's psyche finds contentment in symmetry and repetition, harmonic tremor—whose seismic waveform appears as the nearly perfect repetition of an ocean wave, over and over again for minutes to hours to days—must be highly soothing. However, at Mingo's house in Volcano, harmonic tremor was about as welcome as rabies or a 30-foot python.

The house is about 5 air miles from Mauna Ulu. Rain forest covers this intervening distance, and a 200- to 300-foot-tall ridge bisects it. Thus, we could not see Mauna Ulu in eruption from home, though on nights under a blanket of low clouds, we could see the lava glow reflect off the base of the clouds.

We quickly learned that a distance of 5 miles was close enough for even relatively insensitive humans like us to feel ground tremor from the eruption site when lava fountained high. Mauna Ulu went through twelve such episodes, before slipping into dormancy in 1971; the first was in May and the twelfth in December 1969. We were spared the wavy tremor ride for the first seven simply because we hadn't yet arrived in Hawai'i. Anne, Cinda, and I, though, duly noted episodes eight through twelve. Whether the shaking registered with Mingo, too, we never could tell. Mingo exhibited no overt unusual cat behavior that suggested so, but cats can be truly inscrutable to humans.

Anne and I definitely felt a mild and continuous tremor of the house when Mauna Ulu sent lava fountaining high. This was not a product of our imaginations, nor a side effect of drinking too much 'ōhelo-berry wine. We felt a low-intensity vibration while standing or sitting, but not while walking about. We could not feel the vibration outside the house, no matter how still and concentrated we were. Something in the design of

the house—probably those spindly little wooden posts that supported the house like legs—amplified what little tremor traveled from Mauna Ulu to Volcano. Perhaps a physicist could calculate a coefficient of amplification, knowing the post dimensions, spacing, material properties of the wood, direction to Mauna Ulu, and so forth. I just experienced the tremors.

Cinda sensed the vibration, too, and expressed her awareness of the unusual vibration with atypical dog behavior. She whined just a bit when we could find no other reason for such than the tremor. Mainly, she just looked worried, paced when pacing was not her usual behavior, and seemed generally ill at ease. Doberman critics might say that was just behavior typical of the breed. But those who knew Cinda would disagree. Apparently, Cinda was a canine seismometer who did not like that function one bit.

Mingo? Well, he simply ate, slept, purred, peed, and pooped his way through life, be it in quarantine, in his house at Volcano, or on the mainland. No mild shaking of home could disrupt the pleasures in his simple life. Anne, Cinda, and I were envious.

Though we found watching lava fountains a rare treat, Anne and I were happy when, at the end of 1969, Mauna Ulu changed its pattern of behavior from periodic high fountaining to relatively gentle lava circulation and sloshing contained within a crater at the vent. Five episodes seemed enough for us, considering the shake-up at home that accompanied them.

Meanwhile, back at Mauna Ulu, we carried out a few unique experiments when behavior at the vent changed from the periods of high fountaining characteristic of the volcano's first eight months of activity. For the final twenty months of action, a lake of lava circulated and sloshed within the large crater that had developed around the vent. A scientist's longing for more fountains may in part have prompted our first experiment: we tried to trigger eruptions when none occurred naturally.

The crater that had formed around the Mauna Ulu vent changed in size and shape over time. However, during most of our experiments the crater consistently measured about 300 feet long and 150 feet wide. The top of the contained lake of restless lava tended to fluctuate between 50 and 150 feet below the crater rim. This up-and-down fluctuation repeated in fairly regular cycles and seemed to represent the following chain of events.

From the low level, about 150 feet below the crater rim, the thinly crusted surface of the lava lake slowly rose as gases came out of solution from the magma and became trapped beneath the crust. We likened this process to the rising of a loaf of bread as yeast-driven fermentation produces carbon dioxide that inflates the dough, pushing the top of the loaf upward. With experience—or luck—and good timing, the loaf of bread does not collapse, and one ends with a well-shaped and tasty treat. With the lava lake, however, the accumulating gases eventually lifted the crust to its breaking point, and the underlying gases rapidly escaped to the atmosphere. Their vaporous exit triggered subsidence of the top of the lava lake to roughly its preinflated level. This rise and fall repeated faithfully, in cycles that lasted from several to tens of minutes. We informally named this the gas piston effect.

Time we spent watching these ups and downs from the crater rim combined with almost every geologist's inherent urge to throw rocks led to the discovery that we could cause the lava lake to degas and subside "prematurely." To do so, we simply threw pieces of rock onto the lake's thin, rising crust. These projectiles punctured holes in the crust and thereby began the process of degassing and attendant crustal disintegration before it would have happened naturally.

The crust-veneered lava lake in the Mauna Ulu crater—the site of our experiment in trying to trigger lava to overflow from the crater. The uneven texture on the crusted surface of the lake is the result of plate tectonics as described in chapter 21, "The Dance of the Plates."
—Wendell Duffield photo, U.S. Geological Survey

Our realization that we could at least partly control the timing of the gas-piston cycle spawned the idea of actually triggering an eruption, which we defined as any overflow of lava from the crater. We likened the gas-piston activity to the eruption of a common geothermal geyser. A geyser erupts when surrounding hot rocks heat the geyser's contained water to the point that it begins to boil at the surface. This boiling sends some frothy water over the lip of the geyser. As the weight of the boiling water is removed, pressures at deeper levels decrease accordingly, which triggers a downward-propagating chain reaction of boiling to deeper and deeper levels in the geyser conduit. At the surface, boiling water spews out in surging jetlike fountains. When the geyser's eruption depletes the conduit of boiling water, the eruption ceases, new cool water flows in to refill the conduit, and heating begins a new eruption cycle.

People in Iceland, the home of the world's type example of a geyser, have discovered that they can hasten the onset of eruption by adding a bit of detergent to the near-boiling hot water in the geyser's throat. This additive lowers the water's surface tension and permits a premature onset of boiling relative to uncontaminated water.

We convinced ourselves that if people can trigger the eruption of geysers, we should be able to trigger lava overflow from Mauna Ulu's crater by adding a bit of something to the rising lava column as we punctured the crust to release the trapped gases. For our something, we settled on additional lifting energy. We began our experiments by adding small amounts of water vapor and worked incrementally upward to adding the lifting energy from a charge of chemical explosives.

At Mauna Ulu, we had already produced premature "boiling" of the lava by punching a hole in the thin crust. What would happen, we wondered, if we punched the hole with a container of water rather than a piece of cold rock? Might the instantly formed water vapor add to the natural gassy froth and help lift some melt out over the rim of the crater?

To test this possibility, we began by adding thin-walled one-gallon plastic jugs of water. Running this experiment was as easy as standing at the crater rim and throwing in a jug. But despite repeated runs, our hurled jugs of water never produced an apparent effect beyond that of a cold-rock projectile. So we tossed in incrementally larger water containers, reasoning that more water equals more "extra" vapor, which equals more lift for the lava. We threw in up to five-gallon containers with no apparent affect. Realizing that larger-volume containers would be difficult or impossible to throw over the crater rim, we prepared to take a large upward step in adding a source of energy.

We decided to use explosives. We selected the kind seismologists use when they create their own earthquake rather than waiting for nature to shake the ground. However, we were careful to add only a little of this type of energy—after all, we only wanted to create a piddling overflow of lava at the most.

First, we contacted someone licensed to use the explosives. He was enthused to take part in our experiment and agreed to handle all aspects related to the explosive. Next, we calculated the type and size of container needed to deliver the explosives for detonation a few yards into the lava lake. Within reasonable assumptions about the density of the frothy melt beneath the lake's thin crust, we designed a metal container that would sink at a rate controlled by its bulk density relative to that of the frothy lava. Then we constructed a cable tram across the crater, over which we could transport the container and its package of potential energy and then drop it into the center of the lava lake.

The day of the experiment was charged with excitement for us. The team assembled at the crater rim with all necessary ingredients in hand. We lit the explosive's fuse, whose length was appropriate for the time we needed to get the package into the lava lake. We cabled the container with its sputtering fuse out to the center of the crater and activated the release to drop the package. We retreated to what seemed like an appropriately safe distance and waited—and waited, and waited.

We neither heard, saw, nor felt anything that we could relate to our grand experiment. Our equipment and our idea seemed to have disappeared simultaneously down the drain. Oh well, as all trained scientists know, even negative results are useful. They eliminate one possibility from a stable of hypotheses. We suppose that we badly underestimated the amount of extra energy needed to produce a noticeable change in the natural gas-piston action at Mauna Ulu. We chalked up this experience as one of the learning type, and duly noted that Pele may indeed control her Hawaiian volcanoes. She was behaving in an Old Faithful manner, unresponsive to the nettling and meddling experiments of a few geologists.

20

TAINTED LAVA

Our scientific curiosity unfazed by our failure to trigger eruptions from Mauna Ulu's vent crater, we next designed an experiment to track the movement of lava in the shallow subsurface. We intended to unequivocally document the subsurface interconnection of magma erupting simultaneously at two locations in the Mauna Ulu area. The experiment was clearly defined and tightly constrained. Success seemed assured. With careful planning and flawless execution, both readily achievable, we would be able to write the definitive paper where only arm waving had gone before.

For generations, geologists had waxed long and eloquently about volcano plumbing, the system of passageways through which magma travels in the subsurface. But in spite of convincingly written descriptions and elegant three-dimensional diagrams purporting to depict such plumbing, no one had ever proven a specific subsurface connection between separate sites where magma was erupting simultaneously. Mauna Ulu provided what seemed the perfect opportunity to put fact into foregoing plausible ideas.

That a lava lake persisted for so long at Mauna Ulu indicated a connection to active subsurface lava plumbing that was carrying "new," and therefore hot, melt to the lake. A stagnant pond of lava, isolated from a source of new melt, would have frozen to solid rock within a few weeks to months, perhaps after an episode or two of crustal overturn. Then in April 1970, what seemed to be a leak in the local plumbing sprung as eruption from a newly formed fissure on the flank of Mauna Ulu. This set the stage for our grand experiment.

While lava was circulating in the crater at Mauna Ulu, lava was also spewing from a fissure just a few hundred yards to the southwest, on the flanks of the cone itself. Though that fissure did not connect with the Mauna Ulu crater at the surface, it was radial to the crater. Thus, a combination of orientation, geographic proximity, and simultaneous eruption strongly suggested a shallow subsurface linkage between the crater and fissure and magma "sharing" from a common source of melt. We decided to document what seemed like an obvious connection by adding a chemical tracer to the lava in the crater. We reasoned that the tracer would mix into the crater lava and show up at the fissure as a unique chemical fingerprint, unmistakable when encountered.

We ordered a half ton of a mineral called bastnaesite, to be shipped to HVO from the Mountain Pass mine in southern California. This mineral is very rich in the element lanthanum, whereas natural Hawaiian basaltic magma contains almost none. This difference provides ideal and substantial chemical leverage akin to the unmistakable appearance of red dye in otherwise clear water.

We planned simply to dump the bastnaesite into the lava lake in Mauna Ulu crater and then watch for its lanthanum to appear in lava erupting from the nearby fissure to the southwest. Unlike red dye, however, the lanthanum is invisible to all but the instruments in a sophisticated analytical laboratory. We would have to wait for lab results to verify our hypothesis.

We collected several samples of lava as it erupted from the fissure hours to minutes before the great bastnaesite dump. Immediately after the dump, we collected samples from the fissure every twenty minutes for five hours, and then daily for the next several days. Surely vigorously convecting melt could travel a few hundred yards from crater to fissure in a few days or less.

Is this lava chemically tainted? This 4-foot-tall hornito grew near the base of Mauna Ulu shortly after our chemical-tracer experiment. —Wendell Duffield photo, U.S. Geological Survey

To our surprise and disappointment, absolutely no indication of the lanthanum-rich bastnaesite showed in subsequent analyses of these samples. We concluded that either the bastnaesite-tainted melt flushed out of the crater in a direction other than that of the fissure to the southwest, or that when mixed with such a large volume of pristine melt, this tracer became diluted beyond detection by even the most sophisticated techniques of analysis. Either way, part of me believes that Pele again slapped our wrists for attempting to tamper with her volcano. We had polluted her magma, and she was not about to erupt it back to fulfill our plan.

The fact remains, though, that 1,000 pounds of lanthanum-rich bastnaesite is somewhere in Kīlauean rocks. Someday a geologist may sample Kīlauean basalt accidentally rich with this contaminant. If he or she is aware of our failed experiment, the explanation for such a chemical aberration should be straightforward and simple. Otherwise, I predict an elegantly written tome about chemical heterogeneity within the mantle-source region for magma that rises into Kīlauea, based on an anomalous concentration of lanthanum (and a few other related chemical constituents) in one or two samples. Our experiment may have created the geologic Kensington Runestone of Hawai'i.

21
THE DANCE OF THE PLATES

My next experiment at Mauna Ulu was one of simple visual observations and their interpretation, rather than an attempt to somehow control and modify the volcano. Adding to my belief that we geologists advance our knowledge of Kīlauea mainly at Pele's whim, this passive mode of research taught me a far more significant lesson in volcanology and earth science than all our foregoing attempts at volcano manipulation. This realization dawned on me during one of those many days of watching the convective churnings of the lava-lake pot in Mauna Ulu's crater. But first, let's look at some background information to best understand this part of the Mauna Ulu story.

The late 1960s and early 1970s brought substantial turmoil and revolution to the earth sciences. Old, long-standing ideas fell to the challenge of new and fundamentally different views of how our planet behaves. A serendipitous combination of factors strongly supported the challenge: new technologies and brilliant researchers in such disparate places as cluttered and dusty laboratories at the University of California, Berkeley campus; at Princeton and Columbia Universities; at the U.S. Geological Survey in Menlo Park, California; and on oceanographic research vessels cruising the high seas from domestic and foreign ports. From these sources, a host of new results converged to verify a hypothesis that researchers had floated decades earlier but that had never received even narrow acceptance within the earth sciences. That hypothesis rather suddenly carried the mantle of "ruling paradigm." Originally called continental drift, it soon became global plate tectonics to accommodate seafloors into the model.

From today's perspective, it is difficult to understand why plate tectonics did not take center stage earlier. Now, barely a thinking earth scientist exists who doubts that the earth's crust—or lithosphere, to be technically correct—consists of about a dozen pieces in constant lateral movement relative to each other. The pieces, or plates, move at the rate of an inch or two per year. This "mobilist" model of planet Earth has even gained wide acceptance and understanding in the popular mind, thanks to educational articles and animations in the popular media. Mention of plate tectonics in the least likely social settings commonly draws the sounds of comfortable familiarity with the subject and more than passing interest from guests.

During the late 1960s, though, the word had not yet even begun to spread to the popular audience, and individual geologists were struggling to decide which side of the idea fence was correct for them, mobilist or stabilist. With the passing of just a couple of years, a rapidly increasing number of geologists embraced plate tectonics, particularly when the compelling data of seafloor spreading appeared. Some doubters clung stubbornly to the old view that had held sway for so many generations that it must still be true: the crust may go up and down a bit but it never glides around laterally! And the usual cadre of mugwumps sat with mug on one side of the fence and wump on the other, presumably waiting to position the entire body one way or the other when the correct choice became obvious and a commitment to that choice was intellectually safe.

These times were simultaneously enervating, frightening, exciting, and frustrating for geologists. Perhaps most important, though, they offered great opportunities for major advances in the earth sciences. As a young novice researcher looking for a star to follow and a stable place with which to identify, I was surprised and shocked to see many senior scientists of hero and mentor caliber lose their tempers and even their inner composure when the new theories seemed suddenly to bring their life's work into question. But through all the din of controversy and argument, I began to recognize that enlightened geologists of all ages saw opportunity, not threat, in the paradigm swing. As best I could judge from my inexperienced perspective, the pre-plate-tectonic work did not lack quantity or quality. But we could now reinterpret it in ways that led to a single logical synthesis of many types of data, where multiple, disparate, and lonely explanations once dwelled. Occam's Razor was at work, slicing through the haze of earlier confusion to find the simplest solution to the problem.

I was fortunate enough to be born at a time that put me, as a research geologist, right in the chronological middle of this ongoing revolution. And the crust on the lava lake at Mauna Ulu managed to immerse me in a pool of research that helped test the validity of the new plate tectonic paradigm.

The crust rarely covered the lava lake in one continuous skin. Rather, it was a collage of pieces that moved slowly but noticeably relative to each other. The truly exciting aspect of this restlessness, once I realized what was happening, was that the pieces of crust on the lava lake mimicked those of the entire earth. Day after day, Mauna Ulu featured an ongoing performance of global plate tectonics in miniature. Fortunately for me, tickets were free, and there were enough repeat performances to allow me to document the principal acts of this show.

As with the plates covering our planet, the pieces of lava crust met in three fundamental types of boundary structures: (1) axes of spreading, across which adjacent crustal plates grew as they moved away from each other, (2) axes of collision, across which adjacent plates crashed, ground, and crunched together, folding downward to disappear into the molten stuff below, and (3) axes along which adjacent plates slid laterally past each other. At the whole-planet scale, these are called spreading centers, collision/subduction zones, and transform faults, respectively.

As time between leveling, geodimetering, and other assorted HVO tasks permitted, I visited the lava lake at Mauna Ulu and began to take notes and photographs. I measured and recorded such things as the size range of plates, the maximum thickness of plates, and the speed at which they moved about. Such work was not easy because the plates were about 75 feet below me floating on lava that radiated so much heat that I could look directly over the rim of the crater only in snippets less than a minute long. Thus, I could estimate only a few dimensions with each peek. Still,

Mobile plates of crust that veneered the surface of the lava lake in the Mauna Ulu crater commonly mimicked the motions of global plate tectonics. Zones of orange melt exposed between plates are about 1 yard wide. —Wendell Duffield photos, U.S. Geological Survey

A single large plate (near side) *and a mosaic of smaller plates make contact along a zone of collision, or subduction, that extends nearly all the way across the surface of the lake.*

Two centers of crustal spreading are connected by a transform fault.

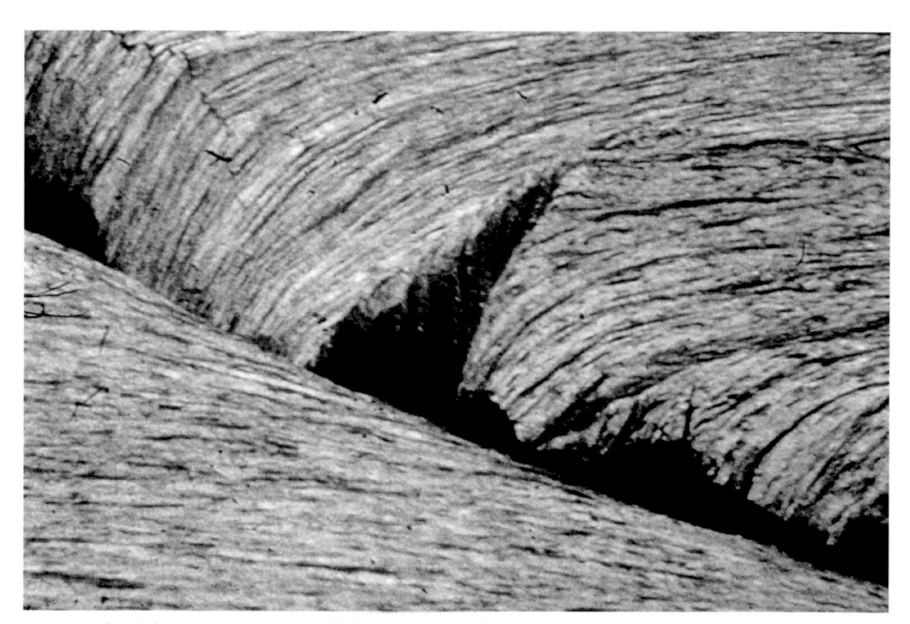

A 3-yard-wide zone of plate collision mimics the sea floor trench part of a subduction zone.

Three centers of crustal spreading share a triple junction.

repeated observations on many different days defined fairly small ranges for all these physical properties.

I exposed many rolls of 16-millimeter motion film—this was before video cameras—to capture the plate motions at true speed. The highly instructive aspect was that the lava-crust plates, unlike the plates that cover our Earth, moved fast enough for the unaided human eye to detect. I supplemented the motion pictures with tens of still snapshots, framed to illustrate each of the three fundamental interplate structures as well as patterns of plates over an entire lava lake.

With a robust set of data in hand, I later calculated that I could reasonably interpret the size, thickness, and speed of lava-lake plates as a scale-model of the whole-earth system. Pele and Mauna Ulu were providing much more than a curiosity. Moreover, professors around the globe were starved for teaching materials to illustrate global plate tectonics to their students, and my still and motion pictures helped satisfy that appetite. I sold hundreds of copies of a twelve-minute movie that I assembled from hours of original footage, and even more copies of a set of 35-millimeter slides that illustrate the three types of interplate structures. By watching the movies, students could grasp the function of a transform fault, whereas study of an endless number of maps of the San Andreas fault—perhaps the world's best-known example—tended to yield only continued confusion.

Global plate tectonics and the version of it at Mauna Ulu became such hot news during the early 1970s that Walter Sullivan, noted science writer for *The New York Times*, described my work in a newspaper story. Sullivan remarked that Mark Twain before me had described plate-tectonic-like activity he had seen in crust on a lava lake at Halema'uma'u Crater. I felt not the least bit scooped by Twain and instead felt flattered to share a mention with him in a story by such a talented contemporary science writer.

The earth science community and I probably owe credit for the discovery of lava-lake plate tectonics to Pele. She undoubtedly had been providing miniature versions of global plate tectonics for millennia, as crusted-over yet circulating lava lakes in craters of Hawaiian volcanoes carried out their dance of the plates. We humans were just slow to recognize what she so readily shared.

22

WHY I CRUISE PARKING LOTS

Starting in June 1971, the lava lake at Mauna Ulu began to decrease in size and in vigor of circulation. The gas-piston behavior, and our attempts to alter it, disappeared from the scene. The dance of the crustal plates ended. Simultaneously, our leveling surveys showed that the summit of Kīlauea began to inflate. Apparently, the magma connection between Mauna Ulu and the summit's magma reservoir was severed, so that magma rising from its mantle source had to be stored beneath the summit rather than erupting at Mauna Ulu. The lava lake completely disappeared in October 1971.

This isolation from a continuous source of new magma persisted at Mauna Ulu until February 1972, when a lava lake reappeared there. Meanwhile, the summit reservoir became so inflated that it leaked magma in two eruptions that originated in Kīlauea Caldera, one in August and one in September 1971. These were both small volume and short lived, probably just spurts of pressure relief along a path to reactivate a connection with

Mauna Ulu. Though small in volume and duration, the September eruption influenced my life in a way that still expresses itself today, about three decades later. This adventure began when I drove one of the HVO trucks to observe the eruption at close range.

The September eruption began along a fissure on the floor of the caldera, between the west wall of Halema'uma'u and the west wall of the caldera, and on line with the adjacent part of the southwest rift zone. From HVO, the quickest and easiest way to get to a close-yet-safe observation point was to drive along Crater Rim Drive to its intersection with this rift zone. There, sediment-floored gaping fissures of the rift zone formed ready-made parking slots in which to get the truck off the main road, out of the flow of any other traffic.

So, I did what seemed quite logical. I parked in one of these slots—but only very briefly because an inner voice carried out a debate whose conclusion was that this parking spot was not ideal. On the one hand, the southwest rift zone had not erupted since 1920, and it had erupted only twice before that in recorded history. This record made extension of the currently erupting fissure into the southwest rift zone seem very unlikely. On the other hand, the active fissure lay exactly on trend with the rift zone. Its southwest end was at the base of the nearby part of the caldera wall, no more than a quarter of a mile from

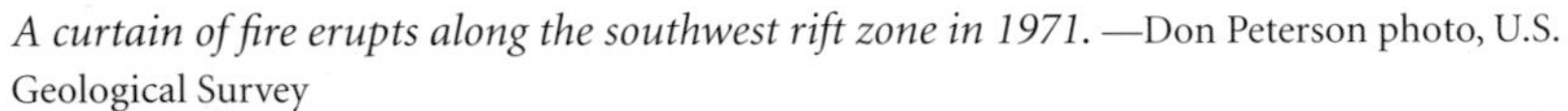

A curtain of fire erupts along the southwest rift zone in 1971. —Don Peterson photo, U.S. Geological Survey

my parking spot. In the end, I moved the truck out of the fissure slot onto the Crater Rim Drive outside the rift zone before walking to the caldera rim to observe the eruption. As it turned out, that was a wise move, although I think it resulted from instinct rather than logic.

Within about thirty minutes, lava began spewing from the "ancient" fissure where I had originally parked. Had I not moved the truck, it would

New, shiny-surfaced lava from the September 1971 eruption (below) *nearly fills the ill-fated parking spot where a pre-eruption, trenchlike slot seemed a logical place to leave the HVO truck. The reprieved truck witnesses the eruption from a safe distance* (right). —Don Peterson photos, U.S. Geological Survey

have been nothing more than a bizarre fossil, a hunk of twentieth-century metal, glass, and rubber buried by basaltic lava. From this modest extension of the erupting fissure, propagating fissures and the eruption migrated down the southwest rift zone some 7.5 miles during the next thirty-five hours. What began as a small eruption in the caldera evolved into an extensive rift-zone eruption. It was only the fourth such eruption there in the past two hundred years, whereas the east rift zone has experienced more than twenty during that same period.

Now I fast forward to my life on the mainland, following those wonderful Hawaiian years.

I can't remember when she first commented, but soon after moving back to the mainland, Anne noticed that I cruise parking lots rather than slip into the first available space. At first I thought she was jesting, but I soon realized she was right. I next went through the rationalization that I was searching for a safe place to park, a slot where other vehicles were least likely to bump our car or slam it with a widely opened door. There may be some truth to that idea, but experience has shown it to be less than a universal truth. I'm convinced that another fuzzy factor is at play in this behavior.

During the more than thirty years since I left Hawai'i, I have never successfully broken my parking-lot peccadillo, even though I have consciously tried more than once. One time, it became a source of some family discord—I drove around shopping for a parking space while Anne wanted to park and shop for merchandise—but it has evolved into a family joke. The point is, it is still there, lurking in my subconscious.

I think I suffer from lingering eruption-parking syndrome, an aftereffect of what happened during that 1971 eruption. I attribute this to Pele, and by now the persistent reader knows how very susceptible I am to Pele's influence. The fact is, though, I never cruised parking lots before that 1971 eruption experience.

23
THE SOUTH FLANK STORY

Time passed quickly for me at HVO. I never wondered what project to start next or what to do on any particular day. Rather, the rest of the HVO staff and I seemed to lack sufficient time to keep track of the restless volcano and, thus, try to forecast what might come next. Quiet moments for planning research strategies and writing up results of recently completed studies came in places other than the office. I found time for reflection mainly in that marvelous claw-foot bathtub in the house that Mingo built, while I soaked away the soreness of another day of leveling or geodimetering, up to my nose in the froth of a bubble bath.

The occasional flatulent burst, intermixed with the commercial lilac-scented bubbles, pungently reminded me that gassy old Mauna Ulu continued to erupt sporadically and at least semipredictably. We were all helping to keep tabs through frequent visits to the vent, daily evaluation of earthquake activity, and periodic leveling, tilt, and Geodimeter surveys. The day-to-day

situation seemed stable enough for us to consider trying some new approach to understanding Kīlauea.

So, as time permitted, Don Swanson and I began to plan leveling and Geodimeter surveys that would encompass the entire south flank of Kīlauea, extending across the grand staircase of landslides that steps down to the south coast. This was to be a unique and very ambitious effort. Our objectives were twofold. We hoped to discover what deformation, if any, had occurred there since previous land surveys, some of which HVO staff and others had completed several decades earlier. We also would look for what that pattern of deformation might tell us about how the volcano behaves away from the summit caldera and its immediate surroundings.

As an additional part of our plan to extend studies away from the summit area, I began to map faults and fissures across the upslope edge of the great south flank, in an area known as the Koaʻe fault system. I did this project solo on the odd "free" days between the various team efforts and during partial days when equipment was under repair or weather conditions were unsuitable for our group surveys.

In my opinion, the greatest achievement (if indeed there is one) of my three years of research at Kīlauea is not my part in the results of measuring and interpreting ground deformation near the summit; not the multitude of observations and interpretations of Mauna Ulu's eruptions; not the successful and failed experiments at Mauna Ulu's lava lake; not the recognition of Pele's version of plate tectonics; not the various studies of the summit and southwest rift zone eruptions in 1970 and 1971; and not completion of my excruciatingly detailed map of geological structures in the Koaʻe fault system. Nor do I count the sum total of all these studies as the jewel in my HVO volcanology crown. Rather, my precious stone sparkles in the results that Don and I, with the rest of the HVO supporting cast, garnered from our leveling and geodimetering of the entire south flank of the volcano. Don served as the scientific, energetic, and spiritual leader of this effort. The rest of us followed, albeit willingly, enthusiastically, and ably.

Though we couldn't have known it as we planned the new surveys, our eventual results helped revolutionize thinking about Kīlauea and its volcano cousins around the world. The results ushered in a couple of decades of intense research that demonstrated that a large volcano can behave like a pile of sand or loose blocks that partly sloughs away in huge landslides when the pile gets too steep or an earthquake shakes it too violently.

Collectively, our surveys of the south flank were a Herculean task. Geography alone made the task much more difficult and time consuming than our periodic and routine geodetic measurements of the summit area. Our team was literally spread thinly over a very large and difficult piece of volcanic real estate.

To compare our results to earlier land surveys, most of which predated the existence of Geodimeters, we had to locate benchmarks that rain forest vegetation had long ago grown over, advancing human developments had bulldozed, or Pele's own lava flows had buried. These efforts burned many calories and subtracted pounds from already lean bodies, as we stumbled through fern, grass, bamboo, and ʻōhiʻa forest, typically advancing only with the help of a savagely swung machete or sugar-cane knife. More than once, an exhausted worker temporarily lay frustrated and immobile in a tangle of fern, unable to move until regenerating sufficient internal energy. And more than once as I bushwhacked my way through densely vegetated terrain, I recalled Cinda's discovery of the reticulated python in Volcano. But we persevered, rediscovered almost all the stations on our "to do" list, and established footpaths to each, never once feeling the crushing coils of a snake.

We at first thought the H.T.S. label (inset) *on some survey stations* (left) *was the initials of one of the pioneering geologists in Hawai'i, Harold T. Stearns. However, the acronym stands for Hawaiian Territorial Surveys.* —Don Swanson photos, U.S. Geological Survey

Initially, when we rediscovered benchmarks boldly stamped with the letters H.T.S., we reveled in the idea that we were following the footsteps of Harold T. Stearns. Stearns and his sometime coworker Gordon Macdonald, who served as HVO's third director, mapped the geology of most of the Hawaiian Islands, including the Big Island, about a quarter of a century earlier. We thought that we had stumbled onto some historic evidence of their pioneering work. However, we soon learned that the H.T.S. letters, while indeed an indicator of old times, stood for Hawaiian Territorial Surveys, not the grand old man in our thoughts.

Once we located the entire network of benchmarks and made them accessible, we had to successfully occupy these reference points—after reaching them on foot and rarely by helicopter—with Geodimeter, reflector, leveling gun, and rod. Moreover, as usual, we had to outwit or outwait the weather so we could complete our measurements within a reasonable time frame—reasonable in the sense that we really needed to finish our surveys before any substantial ground deformation intervened. In the end, it took us months to complete our resurvey of the south flank. We perhaps cheated Pele and the odds by doing so within the framework of many up, down, and sideways motions of the summit area, but motions so local in their effects that the south flank remained an essentially motionless block. We had a wonderfully stable laboratory for accurate geodetic characterization.

Then came the fun and excitement: the comparison of our results with those of the half dozen or so earlier surveys, some of which dated to the late 1800s. The pertinent questions were quite simple. For example, were benchmarks Y and Z the same lateral distance apart in 1896, 1914, 1958, and so forth as in 1970? And had their elevations changed in the interim? An interesting and unexpected challenge reared its head as we began to attempt such comparisons.

Our comparisons of elevations at different times were straightforward. Leveling techniques had improved very little in the intervening years, and as accurately as we could measure, the "absolute" reference of sea level was constant. However, comparisons of horizontal distances came with a fascinating twist.

Land surveyors, not volcanologists, had gathered most of the earlier measurements. And as any concerned property owner knows, such surveyors strive to prove that one's property boundaries have not moved, that the four corners of a rectangular lot are just where they were in the preceding survey. In practice, a relatively small error inherent in all land surveys leads to minor apparent changes of property boundaries, but these typically are so small that they rarely interest even attorneys. So, when their data suggested real, instead of within-error, lateral displacements of benchmarks between each survey, the earlier surveyors chose to distribute this "apparent error" evenly throughout the resurveyed area by mathematical manipulation rather than interpret the changes in the milieu of a shifting and restless volcano. We had anticipated real and measurable ground displacements from our geologic perspective, and this situation forced us to ditch the published and mathematically manipulated results and dig back into the raw data of earlier surveys.

Our final results fit our expectations and were surprisingly consistent with geological data and intuition. And many of our calculated ground displacements were stunningly huge!

Without data from geodetic surveys, the geologic framework alone argues for a generally southward movement of Kīlauea's south flank as the volcano grows. For example, the gaping jigsaw-puzzle fissures abundant within the Koaʻe fault system and rift zones indicate opening along a south-southeast to north-northwest trend. Moreover, Kīlauea grows upon and leans against the south-sloping flank of Mauna Loa, the neighboring volcano that rises nearly three-and-one-half times the elevation of Kīlauea. Simple geometry and mechanics combine to suggest that any lateral motion of Kīlauea should be away from the huge and massive buttress and foundation of Mauna Loa and in the direction of open fissures to the south-southeast. And that is what we discovered.

We could compare survey data back to 1896, nearly seventy-five years before our work. In general, and as expected, we found that horizontal movement toward the south-southeast was the rule and that the amount of movement increased through time. We calculated a maximum horizontal movement of nearly 14.5 feet for land along the south coast near the ominous Halapē campsite that Dallas, Beverly, Anne, and I had occupied. No landowner would be pleased to learn that his neighbor's fence extended almost 15 feet onto his property. Of course, when everyone's land moves by about the same amount, no one loses any land surface area. Still, "new" surface area has been created somewhere, which set me contemplating the status of ownership for the new real estate. This is the stuff of litigation, though—most definitely not a topic for this book.

In general but not entirely as we expected, vertical movements were mostly upward. The uplift was greatest along the rift zones, where magma that intruded sites of eruption apparently lifted adjacent ground. However, the grand-staircase configuration of the south flank graphically illustrates that major downward ground movements, not uplift, are characteristic over the long haul. This observation led us to conclude that our results are atypical of long-term change in land elevation for the south flank. It also prompted us to pay special attention to one particular Geodimeter line that spans the grand staircase along the central part of the south flank.

This roughly 3.4-mile line crosses the central part of the staircase at about a right angle. HVO staff established the survey line in 1965 to provide a monitor of activity for this part of the south flank. Using benchmarks along the shoulder of the Chain of Craters Road assured ease of repeat measurement.

Eleven measurements from 1965 to early 1970 documented 13 inches of shortening. Thereafter, eighteen additional measurements—up to late

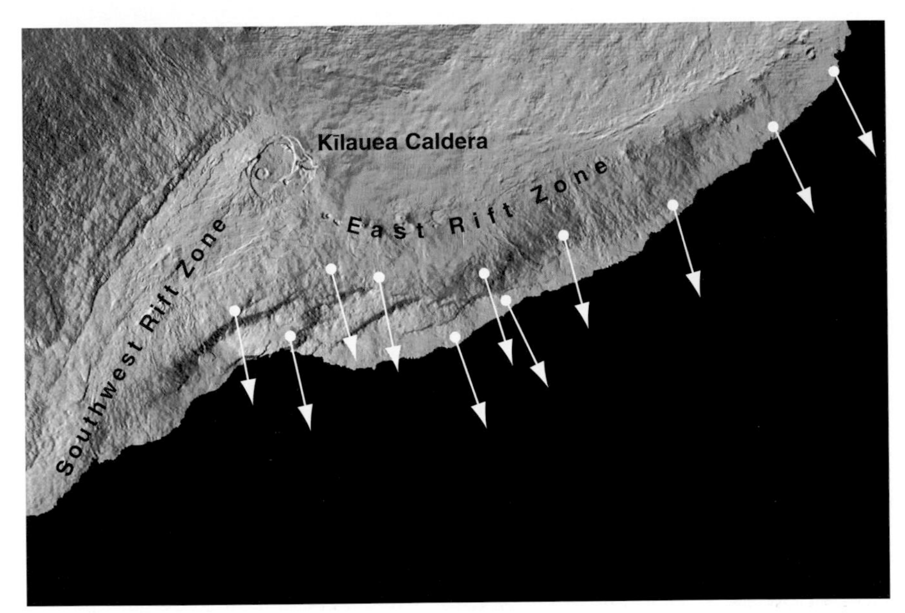

The arrows on this shaded relief image show the direction of lateral ground movements that our Geodimeter surveys documented. —DEM created by Bob Marks, U.S. Geological Survey

1970 when lava flows from Mauna Ulu built up enough to obscure the line of sight—defined a sawtooth pattern of minor changes: repeated shortening and then lengthening. The times of lengthening along this line correlate with shaking by local earthquakes and the times of shortening with newly initiated eruption on nearby parts of the east rift zone.

We thought it possible that earthquake shaking widened gaping cracks and faults that permeate the system of landslide blocks of the grand staircase, whereas intrusion of magma to eruption sites pushed against the same fissures, causing them to contract. On this basis, we speculated that the sawtooth pattern meant the south flank was in a state of metastability, possibly poised for a huge landslide of the type that may characterize the long-term behavior.

We wrote the words expressing this idea in 1973, as part of a professional paper. Nearly two years later, a major landslide rearranged the grand staircase in response to a powerful local earthquake. The formal public release of our forecast had to wait until 1976, about a year after the actual event, when our earlier writings finally appeared as U.S. Geological Survey Professional Paper 963.

Several-hundred-foot-tall risers of the grand staircase are obvious manifestations of headwalls of large landslide blocks on the south flank of Kīlauea. Views parallel (top) *and straight on* (bottom) *show headwalls draped by lavas that the next episode of landslide movement will break.*
—Robin Holcomb photos, U.S. Geological Survey

Our forecast, though speculative, may have been the correct interpretation of just one of many indicators of the current landslide potential of Kīlauea. A contrary view is that our forecast was right for the wrong reason. While we can't prove the behavior of that one Geodimeter line represented the pulse of the entire south flank, we can reasonably suggest a cause-and-effect link. In the spirit of the scientific method, researchers should perform additional tests. Unfortunately, they may have to wait a half century or longer to complete the experiment.

Two earlier earthquakes, in 1823 and 1868, that triggered major landslides on the south flank preceded the 1975 example. However, HVO did not exist then, and the geodetic data that document these are meager indeed. Other scientists before us had speculated on the south flank's structure in ways consistent with the "grand staircase" model. Our new and robust set of geodetic data removed the substantial uncertainty in the finer points of the model while documenting large horizontal movements for the first time.

Since our work, many others have added pertinent information. It all points to a history of periodic reactivation of huge landslides that result from the inherent instability and mobility of Kīlauea's steep south flank. As one of our colleagues has said, you can pile it only so high before the slope becomes unstable. Every child at play who has built piles of sand has seen how the flanks of the pile slip downward when the slope exceeds that critical angle of repose at which it becomes unstable. Throw in the occasional earthquake for good measure and the pile may never even reach its angle of repose. True, a volcano like Kīlauea is not a pile of sand, but its inherent internal strength and penchant for failure may well mimic those of coarse granular debris.

As if the sand analogy doesn't suggest more than enough instability in the volcanic construct of the Big Island, let me introduce an additional penchant for slipping and sliding: the foundation of almost all oceanic volcanic islands is a thick pad of water-saturated mud. On this kind of foundation, the volcanic pile is floating over, rather than pinned to, the underlying sea floor. Imagine your house not firmly tied to underlying soil and bedrock but sitting directly on a pad of slippery ice, and you have an image of how firmly Pele's house anchors to the floor of the Pacific Ocean. With such construction practices of our Earth, it is no wonder that the Big Island and other volcanic islands tend to fall apart from time to time.

Research since the 1970s has produced new and detailed topographic maps of the seafloor around the Big Island and many other geologically young volcanic islands. Those maps show huge, lobate expanses of rough-and-tumble, hummocky terrain that characterize landslide surfaces. Whether our study of Kīlauea's south flank hastened the completion of such complementary research, the timing suggests that our ideas presaged a host of these and other pertinent studies.

The recognition that Kīlauea has a long history of instability and landslides marked the onset of related studies by many researchers at other volcanoes in the sea. For instance, a few years after I left HVO, I successfully applied the Kīlauea-falls-apart model to La Réunion Island in the Indian Ocean. The fundamental pattern of similar behavior is clear, and it makes the south flank story one of broad application.

24
MINGO GOES HOME

My three years at HVO seemed to end almost before they began. I was always too busy there to think much about the passage of time—and I was having too much fun. By August 1972, it was time to move on and make way for another scientist from the mainland to take my place, just as Don Swanson and Dallas had done a few months earlier. My training was complete enough to cycle another volcano novice into the HVO school of hands-on experiences.

In three short years, I had transformed from a wet-behind-the-ears Ph.D. who had studied only granite to a fairly seasoned volcanologist. I had experienced the pains and pleasures of geodimetering and leveling and learned to recognize some of Kīlauea's secrets hidden in the geodetic stories. I had ridden out hours of harmonic tremor and learned to appreciate, if not enjoy, the significance of Pele's magma-energized one note. I had been pelted and a little bit melted by hot cinder fallout from Mauna

Ulu fountains and had learned to respect the power of active volcanoes from this and other breathtaking experiences. I had delighted in a clear and global vision of lava-lake plate tectonics while most of my colleagues battled the befuddlement of studying small bits and pieces of the whole-earth version. And I had contributed to significant advance in understanding how an entire volcano, not just the summit area, behaves. With the vision of hindsight, I know that the three HVO years shaped most of the rest of my career in geology.

Departure day for Anne, Mingo, Cinda, and me was three years and a month after Neil Armstrong's moonwalk and three years after our version of that great travel adventure. First came a typical Big Island lūʻau, filled with tasty Hawaiian food and sad good-byes. We probably all ate and drank too much, but Pele cooperated by not staging an eruption that would have demanded our attentions while we recovered from overindulgence. In saying good-bye, I did my best to express thanks to the permanent HVO staff, the employees drawn from the local population. With the continuity and day-in-day-out support they provide, each research scientist who visits from the mainland for just a few years can accomplish much. Without the locally recruited staff, the business and breadth of HVO research would have greatly diminished or collapsed long ago. Jaggar recognized this truth early in the life of HVO.

When the party was over, Anne and I got down to the business of once again moving our household. The reverse process was even easier than the original move. We still had very little material property. Our mainland burglar had not followed us to Hawaiʻi, but we bought nothing of value or bulk in the Islands. We sold the Pontiac—whose resale value was less than the cost of shipping the car back to the mainland—to John, the HVO machinist. Both Mingo and Cinda would travel with Anne and me, Cinda in the cargo hold where large pets go and Mingo with us in a travel box that fit under an airplane seat.

We prepared Mingo for the flight "home" by visiting our Hilo veterinarian to discuss travel procedures for cats. He suggested administering a tranquilizer in hopes of avoiding or at least minimizing any flight-induced trauma. To test the effects of the tranquilizer and reduce the possibility of in-flight surprises, we fed Mingo his prescribed dose of medicine a few times at his house in Volcano some days before our departure. These practice sessions produced a perfectly drowsy and tractable animal. However, practice sessions proved totally unlike the real travel experience.

The flight east from Hilo began quietly. Both Anne and I shed a few private and shared tears in sadness of ending our Hawaiian adventure. But we knew this day was coming, even before we had made the westward trip to the Big Island and our Volcano home. From his in-flight behavior, we concluded that Mingo was equally sad about leaving the tropical paradise, but he had not known that the Hawaiian sojourn had to be limited.

About an hour into the five-hour flight, Mingo became agitated, tranquilizer notwithstanding. Fortunately for the other human passengers, he expressed his agitation by physical thrashing about, rather than

caterwauling. Anne and I first noticed his displeasure when scratching sounds emanated from inside his airline-approved cardboard travel box. This behavior seemed harmless enough, until Mingo's hind claws and his teeth began to appear through the walls of the box. We were sure the front claws would have appeared, too, if he still had them. Unless we reinforced that carrier, he would literally claw and chew his way out. And although he was by airline policy the only cat in the human-passenger compartment, the vision of Mingo streaking free and in terror about the cabin was not a pleasant one.

The flight attendant suggested tape reinforcement, and quickly fetched a roll borrowed from the flight engineer. We covered the tooth-and-claw holes and waited. Those ivory-colored talons soon reappeared, and the tape was refetched. After three or four repeats of fetch and patch, the flight attendant rather grudgingly gave us the entire roll of tape to keep and use as needed. Much to our relief, this was sufficient to keep Mingo contained for the rest of the flight. But to our dismay, the experience left us wondering why a roll of tape is so critical to the success of a flight engineer's duties. Does he keep baling wire on hand, too? Perhaps we simply misinterpreted the attendant's hesitancy to let us use the entire roll in the first place.

In the end, Mingo relaxed back to a tranquil, if not tranquilized, state for the last hour or so of the flight. Though he was unaware of his influence, we felt he had successfully led the family west to the Hawaiian paradise, and intended or not, his displeasure of the eastward flight mirrored feelings Anne and I shared about returning to the mainland. But life goes on.

At fourteen years, Mingo succumbed to old age. We immediately replaced him with a virtual physical clone, another black alley cat with emerald green eyes. We named the replacement Déjà Vu, in recognition of the resemblance. Déjà lived with and carried us through many family adventures during the ensuing years, and he always served as a visual reminder of Mingo and our modest step for volcanology.

25 EPILOGUE: SHARING THE FRUITS OF MATURATION

Many of you vividly remember when Mount St. Helens, a towering strato-volcano near Vancouver, Washington, erupted violently in May 1980. That eruption leveled nearly 200 square miles of adjacent Douglas-fir forest and blanketed ash over terrain hundreds of miles downwind to the east. Such volcanic violence was very unusual in the lower forty-eight states, at least since European explorers and settlers entered the West; until St. Helens blew, only Lassen Peak, in a remote part of northern California, had erupted during the twentieth century. But unlike Lassen, St. Helens borders the densely settled metropolitan areas of Vancouver, Washington, and Portland, Oregon. A several-month period encompassing the eruption was a worrisome time for politicians, businessmen, geologists, and the general public.

At that time, my stint at HVO was eight years behind me, though it remained close to my heart. Early in 1980, I was submersed in paperwork associated with coordinating the U.S. Geological Survey's Geothermal Research Program. A team was readying a newly established system of seismometers as part of a broader effort to assess the geothermal-energy potential of the Cascade Range, a chain of volcanoes that extends from southwestern Canada through western Washington and Oregon and into northern California. Ironically, the seismic network became functional just a few days before Mount St. Helens reawakened on March 20, 1980. That day, many earthquakes, all centered directly below the volcano, shook the mountain and the surrounding area, as magma began to rise toward the surface.

Thus, the Geothermal Research Program and its new seismic network were predecessor and midwife to the Cascades Volcano Observatory, or CVO, the U.S. Geological Survey's second formal volcano observatory. CVO grew from the need to continue to watch Mount St. Helens and other Cascade volcanoes, which are sure to erupt in the not-too-distant (geologic) future.

Because an eruption from Mount St. Helens could potentially destroy nearby densely populated areas, the newly restless behavior of the volcano demanded immediate attention. The details of the Mount St. Helens eruption are not appropriate to this book, and you can read them in U.S. Geological Survey Professional Paper 1250 (see Additional Readings). However, we can largely attribute whatever success we enjoyed in recognizing hazards and mitigating risks from the 1980 eruption of Mount St. Helens to the application of volcano-monitoring techniques developed at HVO and to the instant availability of many competent volcano doctors, most of them trained at HVO. Many of the personnel best prepared for the task were HVO alumni and other U. S. Geological Survey geologists who had been assessing geologic hazards in the Cascade Range. I played a role in this initial fire drill of organization, but from the end of my desk-anchored tether, I could only watch what happened later at Mount St. Helens.

The assembled team of volcanologists began scurrying across and around Mount St. Helens within a few days after the onset of earthquakes. The team's initial study of the volcano included strategies familiar to those on Kīlauea: continued and increased monitoring of seismicity; attempts to measure any ongoing ground deformation, such as bulging, cracking, and subsidence; and plain old visual observations as frequently as feasible. As needed, current HVO staff worked with the HVO alumni and all other scientists, trying desperately to evaluate the potential for Mount St. Helens to erupt. The detailed pattern of events leading up to the eruption differed in many ways from that typical of Kīlauea Volcano. But, tracking volcanic seismicity and ground deformation provided the keys to understanding Mount St. Helens well enough to avert what might have been a colossal disaster in number of human deaths.

In the weeks and days before the eruption, the onsite crew referred to a carefully documented history of eruptions from Mount St. Helens, and they developed potential eruptive scenarios. The earlier work spelled out the timing and character of the volcano's eruptions during the preceding four millennia. Many had the earmarks of violent explosions. The authors, U.S. Geological Survey geologists Dwight Crandell and Donal Mullineaux, had published the results of their geologic sleuthing only two years earlier, in 1978. With a clear foresight often missing in geologic problem solving, they had written, "The volcano's behavior pattern suggests that the current quiet interval will not last as long as a thousand years; instead, an eruption is more likely to occur within the next hundred years, and perhaps even before the end of this century."

On Mount St. Helens, the fruits of a mature Hawaiian observatory were freely and beneficially shared across the Pacific table. As this sharing continued, the U.S. Geological Survey established three more volcano observatories.

In 1988, the Alaska Volcano Observatory, or AVO, was founded to study and forecast eruptions from Alaskan volcanoes. Some of Alaska's volcanoes pose a direct threat to the state's growing population centers. Others extend westward from near Anchorage to the end of the Aleutian Island chain and spew columns of rocky ash directly into the flyways of commercial aircraft. More than one near disaster has occurred when pilots unknowingly flew into ash plumes, and the jet engines ingested this indigestible stuff. The staff at AVO now regularly and promptly broadcasts public notices when such volcano-caused dangers sully the sky.

In 1999, following two decades of informal observatory activities, the U.S. Geological Survey formally established its fourth volcano observatory: the Long Valley Volcano Observatory, or LVO, in east-central California's Long Valley. Scientists at LVO study the young volcanic region known as the Long Valley Caldera. This caldera collapsed in response to the instantaneous eruption of 20 cubic miles of magma about 700,000 years ago. Since then, several small eruptions have occurred within and along the edges of the caldera. The caldera area came back to life in 1980 as once again magma rising toward the surface began producing earthquakes, fissures, ground bulging, and locally high discharges of carbon dioxide gas. The most recent eruption in Long Valley (as of 2002) was only a few hundred years ago, and this unstable situation requires close monitoring, especially in view of the potential harm to local populations and the heavy tourist traffic to the eastern Sierra Nevada.

In similar fashion and for similar reasons, the U.S. Geological Survey established the Yellowstone Volcano Observatory, or YVO, in 2001. Yellowstone Caldera collapsed when 240 cubic miles of magma instantaneously erupted 640,000 years ago. As at Long Valley Caldera, many eruptions have occurred within Yellowstone Caldera, and today earthquakes bulge and shake the ground as the underlying magma becomes restless.

The organization of the four mainland volcano observatories is patterned after HVO but allows for the unique needs of each research center.

Another by-product of the U.S. Geological Survey's volcano program that began at HVO is the creation of a volcano-hazards fast-response team. This small group of U.S. Geological Survey volcano experts stands ready to fill requests for help both at home and abroad. The response team uses CVO as its home base, where volcano-monitoring equipment and personal clothing are ready to be packed at a moment's notice, whenever the team receives a request for assistance. The pages of the team's passports sport the stamps of countries around the Pacific Ring of Fire, that ocean-girdling belt of volcanoes. Team members also commonly teach training courses for countries that want to build a cadre of their own talented volcano experts. Thomas Jaggar would be proud of this group because he viewed the need for volcano monitoring as knowing no international bounds.

Staff members and alumni of HVO were also instrumental in establishing the World Organization of Volcano Observatories, or WOVO, in 1981.

And so it is that a worldwide network of volcano observatories grows. As it does, we pursue Jaggar's vision of an international *Ne plus haustae aut obrutae urbes* ("No more burned or buried cities").

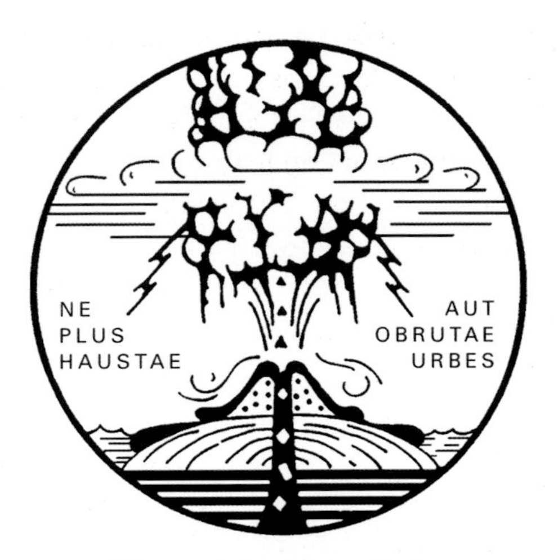

Thomas A. Jaggar's logo for the Hawaiian Volcano Observatory

A GALLERY OF DIRECTORS AND SCIENTISTS-IN-CHARGE OF THE HAWAIIAN VOLCANO OBSERVATORY, 1912–2002

The world of volcanology has benefited substantially from the planned rotation of scientists through HVO. Such rotation exposes a "practical-maximum" number of scientists to training, discovery, and intellectual growth that naturally come with time spent at an active volcano laboratory. It simultaneously exposes the mission of the observatory to the benefits that naturally come with a variety of personalities and abilities. Through staff rotation, HVO serves as a training ground for many volcanologists, while the multitude of new hands and minds unlocks the secrets of Hawaiian volcanoes relatively quickly. —Hawaiian Volcano Observatory archives photos

Thomas A. Jaggar, 1912–40

Ruy H. Finch, 1940–51

Gordon A. Macdonald, 1951–56

Jerry P. Eaton, 1956–58, 1960–61

Kiguma J. Murata, 1958–60

Donald H. Richter, 1961–62

James G. Moore, 1962–64

Howard A. Powers, 1964–70

Donald W. Peterson, 1970–75, 1978–79

Robert I. Tilling, 1975–76

Gordon P. Eaton, 1976–78

Robert W. Decker, 1979–84

Thomas L. Wright, 1984–1991

David A. Clague, 1991–1996

Margaret T. Mangan, 1996–1997

Donald A. Swanson, 1997–present

ADDITIONAL READINGS FOR THE TECHNICALLY INCLINED

Chadwick, William W., and others. 1993 and 1994. Maps of bathymetry around the south and west submarine slopes of the Big Island. United States Geological Survey Maps MF-2231 (1993), 2233 (1993), 2255 (1994), and 2269 (1994).

These maps show hummocky seafloor topography that represents the offshore chaos of landslides from Big Island volcanoes.

Decker, Robert W., Thomas L. Wright, and Peter H. Stauffer, eds. 1987. *Volcanism in Hawai'i.* United States Geological Survey Professional Paper 1350. 1667 pp.

A collection of rather technical papers that describe our knowledge of volcanism in Hawai'i, as of 1987. Published in celebration of the seventy-fifth anniversary of HVO.

Duffield, Wendell A. 1972a. A Naturally Occurring Model of Global Plate Tectonics. *Journal of Geophysical Research* 77 (14): 2543–55.

A technical description and interpretation of lava-lake plate tectonics at Mauna Ulu.

———. 1972b. Kīlauea Volcano Provides a Model for Plate Tectonics. *Geotimes* 17 (4): 19–21.

A brief, popular description of lava-lake plate tectonics at Mauna Ulu.

———. 1975. *Structure and Origin of the Koa'e Fault System, Kīlauea Volcano, Hawai'i.* United States Geological Survey Professional Paper 856. 12 pp.

A technical description and interpretation, with map, of faults, fissures, and pali of the Koa'e fault system.

Duffield, Wendell A., Robert L. Christiansen, Robert Y. Koyanagi, and Donald W. Peterson. 1982a. Storage, Migration and Eruption of Magma at Kīluaea Volcano, Hawaii, 1971–1972. *Journal of Volcanology and Geothermal Research* 13: 273–307.

Technical descriptions and interpretations of ground deformation and eruptions at Kīlauea during 1971 and 1972.

Duffield, Wendell A., and Richard S. Fiske. 1989. "A Teacher's Guide to Questions and Answers and Lab Exercises," accompanying the videocassette *Inside Hawaiian Volcanoes.* United States Geological Survey Open-File Report 89-685. 13 pp.

Teaching materials related to Hawaiian volcanism and plate tectonics. Appropriate for middle school to undergraduate levels.

Duffield, Wendell A., Laurent Stieltjes, and Jacques Varet. 1982b. Huge landslide blocks in the growth of Piton de la Fournaise, La Réunion, and Kīlauea Volcano, Hawaiʻi. *Journal of Volcanology and Geothermal Research* 12: 147–60.

Comparison of the role of huge landslides in the growth of Kīlauea and Piton de la Fournaise Volcanoes.

Eaton, J. P. 1959. A Portable Water-Tube Tiltmeter. *Bulletin of Seismologic Society of America* 49: 301–16.

A detailed description of the made-from-artillery-casings tiltmeter designed by Jerry Eaton and built at HVO.

Eaton, J. P., and K. J. Murata. 1960. How Volcanoes Grow. *Science* 132 (3432): 925–38.

An early summary paper about the structure and balloonlike behavior of Kīlauea Volcano.

Fiske, R. S., T. Simkin, and E. A. Nielson, eds. 1987. *The Volcano Letter.* Washington, D.C.: Smithsonian Institution Press. 530 issues.

A reprinting of 530 issues of an HVO weekly newsletter, originally published between 1925 and 1955.

Kling, G. W., M. A. Clark, H. R. Compton, J. D. Devine, W. C. Evans, A. M. Humphrey, E. J. Koenigsberg, J. P. Lockwood, M. L. Tuttle, and G. N. Wagner. 1987. The 1986 Lake Nyos Gas Disaster in Cameroon, West Africa. *Science* 236: 169–75.

Technical story of a giant "burp" of carbon dioxide from a dormant volcano, which suffocated hundreds of people and their livestock.

Lipman, Peter W., John P. Lockwood, Reginald T. Okamura, Donald A. Swanson, and Kenneth M. Yamashita. 1985. *Ground Deformation Associated with the 1975 Magnitude-7.2 Earthquake and Resulting Changes in Activity of Kīlauea Volcano, Hawaiʻi.* United States Geological Survey Professional Paper 1276. 45 pp.

Technical description and interpretation of reactivation of huge landslide blocks of Kīlauea's grand staircase and the landslide-generated tsunami (tidal wave) that washed across the Halapē area.

Lipman, Peter W., and Donal R. Mullineaux, eds. 1981. *The 1980 Eruptions of Mount St. Helens, Washington.* United States Geological Survey Professional Paper 1250. 844 pp.

Comprehensive description of geologic and human events leading up to, during, and immediately following the violent eruptions of Mount St. Helens in 1980.

Macdonald, Gordon A., Agatin T. Abbott, and Frank L. Peterson. 1983. *Volcanoes in the Sea.* Honolulu: University of Hawaiʻi Press. 517 pp.

A comprehensive and richly illustrated summary of the volcanic geology of the entire chain of Hawaiian Islands.

Swanson, Donald A., Wendell A. Duffield, and Richard S. Fiske. 1976. *Displacement of the South Flank of Kīlauea Volcano: The Result of Forceful Intrusion of Magma into the Rift Zones.* United States Geological Survey Professional Paper 963. 39 pp.

Description and interpretation of geodetic surveys of the entire south flank of Kīlauea. This is the first documentation of the direction and magnitude of lateral movements of this part of the volcano.

Swanson, Donald A., Wendell A. Duffield, Dallas B. Jackson, and Donald W. Peterson. 1979. *Chronological Narrative of the 1969–71 Mauna Ulu Eruption of Kīlauea Volcano, Hawai'i.* United States Geological Survey Professional Paper 1056. 55 pp.

Detailed and richly illustrated description of the eruption that inspired Duffield to write Chasing Lava.

Tilling, Robert I., Christina Heliker, and Thomas L. Wright. 1987. *Eruptions of Hawaiian Volcanoes.* United States Geological Survey. 54 pp.

An extensively illustrated 54-page general-interest booklet that is an excellent source of information about Hawaiian volcanism for those who are not volcanologists.

Ulrich, George. 1991. Report on a Burn Incident, June 12, 1985, East Rift, Kīlauea Volcano. *Earth.* January: 10–14.

A first-hand account of falling (both legs up to the knees) into an active lava flow and living to tell about it.

Wright, Thomas L., Taeko Jane Takahashi, and J. D. Griggs. 1992. *Hawai'i Volcano Watch.* Honolulu: University of Hawai'i Press and Hawai'i Natural History Association. 162 pp.

A pictorial history of Big Island volcanism from 1779 to 1991.

INDEX

ABOUT THE AUTHOR

—Anne Duffield photo

Wendell A. Duffield received a Ph.D. in geology from Stanford University in 1967. During the following three decades, he studied volcanoes around the world as an employee of U. S. Geological Survey. Now retired, he continues to chase lava as a Scientist Emeritus with the U. S. Geological Survey. Duff also teaches geology at Northern Arizona University, consults about geothermal energy, and writes books and magazine and newspaper articles for a general audience. Duff and his wife, Anne, their dog (Pele), and cat (Marza) spend summers in the beautiful lake country of northwestern Wisconsin and winters in mild Flagstaff, Arizona.

In addition to *Chasing Lava*, Mountain Press Publishing Company publishes a series of Roadside Geology guides, Roadside History guides, full-color plant and bird guides, outdoor guides, horse books, and a wide selection of western Americana titles.

For more information about our books, please give us a call at 800-234-5308, or mail us your address and we will happily send you a catalog. If you have a friend who would like to receive our catalog, simply include his or her name and address. Thank you for your interest in our titles and for supporting an independent press devoted to providing high-quality books to readers interested in the world around them.

MOUNTAIN PRESS PUBLISHING COMPANY
P.O. Box 2399 / Missoula, MT 59806
406-728-1900 / TOLL FREE: 800-234-5308 / FAX: 406-728-1635
e-mail: info@mtnpress.com / website: www.mountain-press.com